Ladybirds II

Ladybirds II

The Continuing Story of AmericanWomen in Aviation

Henry M. Holden

and

Captain Lori Griffith

Mt. Freedom, N.J.

First Edition
First Printing - Feb. 1993

Cataloging in Publication Data

Holden, Henry M.
Ladybirds II: the continuing story of American women in aviation/by Henry M. Holden and Lori Griffith.
 p. cm.
 includes bibliographical references and index
 ISBN 1-879630-12-5

1.Women air pilots--United States--History. 2. Women in aeronautics--United States--History. I. Griffith, Lori. II. Title. III. Title: Ladybirds two.

TL521.H58 1993   629.13'092'273
                                 QBI92-20153

Cover photo credit: top left Capt.. Lori Griffith; top right, Alinda Wikert; bottom right, Wally Funk; bottom left, Julie Clark; center, Eileen Collins, NASA

This book is available for purchase by educational institutions and groups at bulk discount prices. Write the publisher for discount information.

Printed on recycled paper

Table of Contents

Ladybirds II
The Continuing Story of American Women in Aviation

Part II

Chapter Four - Daughters of Minerva   **81**

Chapter Five - Sport Aviation   **119**

Introduction

While we briefly cover some of the early pioneers, this book is an ongoing, and expanding record of lives and events unfolding as well as a record of those already finished.

The women celebrated here are from every walk of life who followed their vision and overcame resistance to accomplishing their dream. Since these women have been achieving their goals from the beginning of human flight, this book and its predecessor *Ladybirds - The Untold Story of Women Pilots in America* can only be a bare-bones record of their accomplishments.

For years women in aviation faced incredible obstacles, known or inferred. Social ostracism, scorn and defamation were common. Their accomplishments were often ignored, ridiculed or forgotten. Many of the early pioneers were not recognized and some paid a heavy price for doing what they empowered themselves to do.

We hope the reader will be excited by the stories of the pioneers, inspired by the current women in aviation, and applaud all.

Regrettably there are exclusions to this history of American women in aviation. In selecting women for this book we encountered some obstacles. Since the activities of women in the past, more often than not were considered to be of little importance, many of their accomplishments went unnoticed, or unpublished in books or mass media. We also encountered an attitude on the part of some women themselves that their accomplishments were of no special significance and therefore not worth sharing. Sometimes individuals had been recognized in the media but that recognition took place decades ago, and today there is no memory of their accomplishments.

The planned list of achievements filled this book to its design size and overflowed. Numerous women whose stories and photos deserve to appear here had to be left out because the text had grown too long. So on that note we say there is no such thing as a "complete" list of the accomplishments of American women in aviation. And that is reason for rejoicing.

While we hope this book will be used as a reference, it should be seen equally as an ongoing tribute to the women of the United States, especially the unacclaimed and unknown, living and dead, who have been the pathfinders and on the front line of the vanguard of progress toward equal status, equal treatment and equal place in the history of humankind.

Acknowledgements

It is difficult for us to adequately acknowledge and thank all the people who made this book possible. We have tried to list them all on page 315. If we have omitted anyone it is unintentional.

There are several people who generously volunteered a significant portion of their time to help us make this book a valuable and accurate reference for women in aviation. Among them are Amy Carmien, President of *Women in Aviation*, and Jody Clark. They patiently edited the manuscript and made many constructive comments that improved the quality of the text.

H. Glenn Buffington, also volunteered to read the manuscript for historical accuracy. His comments were of great value and they helped to improve the end product.

Lt. Cmdr. Trish Beckman was enormously helpful by providing access to the women who flew in Desert Storm and was instrumental in helping to get their story into print.

Betty Skelton Frankman and Kay Brick provided both inspiration, special materials and support to the project. Kay also supplied most of the photos to illustrate the story of the Powder Puff Derby, and Transcontinental Air Races.

A special thanks goes to Wally Funk who shared with us her notes, clippings and records of her days as one of America's first astronaut trainees.

And of course we owe a special acknowledgement to Nancy (Holden) and Gregg (Griffith), who supported and understood our need to write this history of American women in aviation. To everyone, our thanks and best wishes.

About the Authors:

This is Henry M. Holden's fifth book. His previous books are, *The Douglas DC-3; The Boeing 247 - The World's First Modern Commercial Airplane; The Fabulous Ford Tri-Motors, and Ladybirds - The Untold Story of Women Pilots in America.* He has published more than 250 magazine and newspaper articles that have appeared in a New Jersey newspaper syndicate, and the *New York Times Syndication.* His articles have appeared in various national magazines including, *USAir, Peterson's Photographic* , and various aviation magazines including the prestigious American Aviation Historical Society's *Journal,* and *Aviation Heritage.* He is a member of the Aviation Hall of Fame of New Jersey, the American Aviation Historical Society, and the Aviation/Space Writers Association. Holden lectures frequently on aviation and business topics and is currently working on his next book, *The Whirly-Girls: International Women Helicopter Pilots.*

Captain Lori Griffith received her private pilot rating when she was 18, and attended Indiana State University on a Talent Grant Scholarship in their aviation program. She began in 1979 and finished a four year degree with a double major in Aviation Administration and Professional Flight in two years.

Immediately following college she was hired by a company in Kentucky where she flew pipeline patrol of oil and gas lines. In August of 1981, she was hired by Atlantis Airlines, a commuter in the Southeast. Because She was so young she could not receive her ATP until her 23rd birthday. A captain's position came available in early December 1983, and both the Chief Pilot and Training Officer flew the line for her to hold the slot until her birthday. On that day, she was the youngest female captain in the industry.

She was hired by Piedmont Airlines in March 1984, as a flight engineer on the Boeing 727 and up-graded to first officer just twelve months later. Two years later she received her captains bid on the Fokker F-28 and became the youngest female jet airline captain at the age of 26.

Captain Griffith has the following ratings; Commercial, Instrument, Multi-Engine, Seaplane, Rotocraft, Glider, Instructor Airplane, Instructor Seaplane, Instructor-Instrument and Instructor-Multi-Engine. Today, Captain Griffith is a Boeing 737 Captain with USAir.

Captain Griffith is an active member of the International Society of Women Airline Pilots organization where she has served as an executive Council member and currently as their Master Seniority List Chairman. She is also the organization's Museum Exhibit Chairman and designs permanent exhibits around the country on women airline pilot history. Her husband, Gregg, is a Boeing 757 first officer for United Parcel Service and together they share their love for flying.

Captain Griffith is also a member of the International Whirly-Girls, the Wings Club and the Ninety-Nines.

Foreword

As the twentieth century draws to a close, women in aviation find themselves at the threshold of a new beginning, rather than at the end of an era. A beginning that exceeds the boundaries our spirited pioneers originally set, and one that yesterday's lofty legends could only dream would one day exist.

Past generations could not conceive that women would command our space flights and protect our skies at war, but that very reality is unfolding as we move towards the year 2000. It is by choice, not by chance that today's women have propelled themselves into virtually every aspect of aviation, and it is their success that inspires others to follow in their footsteps.

As an ongoing tale in the Ladybirds series, *The Continuing Story of American Women* in Aviation is as rich as the heritage from which we hail. It is a glimpse into the lives and diverse career histories of some of the industry's most talented women; from engineers, educators and entrepreneurs to authors, airline CEO's and air safety investigators.

Together these and many other women have succeeded in professions that stretch far beyond the realm of just the cockpit and have overcome opposition and adversity to turn opportunities in aviation into timeless triumphs. No longer compelled to prove their abilities or substantiate their presence, women have made impressive contributions in their fields and continue to be a driving force within the industry.

Only time will tell what milestones will be marked by aviation's women of tomorrow, but if the source of their accomplishments lies in the power of their imagination, there is truly no limit to what they will achieve.

Chapter One

The Dawn of Aviation (1783-1919)

I was annoyed from the start by the attitude of doubt by the spectators that I would never really make the flight. This attitude made me more determined than ever to succeed. Harriet Quimby 1912

Over the past 80 years women have played a significant role in the development of aviation in America. But, women did not first enter aviation a mere 80 years ago. They were contributing from the dawn of flight. It is only in recent years, however, that public awareness has grown around their contributions.

Just seven months after the first man flew, the first woman went aloft. And five months before the Wright brothers first flew their heavier-than-air machine, a woman soloed in a powered aircraft.

Human flight began on November 21, 1783, when Joseph and Etienne Montgolfier, two dashing young Frenchmen, released a hot air balloon of their own design - a lemon-shaped bag made of paper lined with cotton, and held together with buttons. A wicker basket hung beneath the mouth of the bag, and in its center a fire burned in a brazier, fueled by bales of hay and wool, heating the air inside the balloon. Twenty-five minutes later, and five miles away, they landed. They became the first humans to go aloft and return safely.

On December 1, 1783, Professor Jacques Alexander Cesar Charles introduced the first of many innovations in the history of flight, the hydrogen balloon, a rubber coated silk bag filled with hydrogen gas made from sulfuric acid applied to iron.

Madame Thible believed that she could do this gravity-defying feat also. Seven months later, on June 4, 1784 at Lyons, France, Madame Thible went up in a Montgolfier balloon with an artist/pilot named Fleurant. She volunteered for the flight when she heard that Fleurant was having trouble finding a male companion for his proposed flight. She is supposed to have said, "I'll go with him to heaven and to glory." She became according to the French, "The world's first female aeronaut." When she and Fleurant landed she was euphoric and treated Fleurant to an aria from *La Belle Arsene* singing, "I am victorious - I am Queen."[1]

In 1798, Jeanne Labrosse became the first woman to solo a balloon, in France. She would not be remembered or gain a place in history because another woman would gain prominence in those early days of flight. Her name was Marie Blanchard.

Madame Marie Blanchard

Napoleon's air service chief was Madame Marie Madeleine Sophie Blanchard (nee Armant). She owes the distinction of her office to her fame as one of the first balloonists.

There is a romantic legend that before Marie Madeleine Sophie Armant was born, a young Frenchman was passing through the countryside. He came across a pregnant woman toiling in the fields. Her condition and obvious poverty moved the young man. In a moment of whimsy the man said, "If your child is a girl I will come back when she is sixteen and marry her." The man, legend has it, moved on, but returned when young Marie was sixteen, and kept his promise. The man was Jean Blanchard, twenty-five years her senior, and already noted for his famous balloon flight across the English Channel from Dover, England to Calais, France.

Madame Blanchard did not make her first balloon flight until 1804, and the circumstance that led to her becoming Napoleon's chief of the air service, occurred at the coronation of Napoleon and Josephine. Balloons and fireworks figured prominently in Europe in those days. Part of the celebration was a flight of balloons from Rome, to Paris, with one decorated with a large crown bejewelled with colored lanterns. Napoleon's air service chief was in charge of the balloon flight from Rome. Because of strained relations, Napoleon did not want any mishaps on the Italian side of the celebration. Unfortunately, the balloon with the crown ran into trouble and fell swiftly to earth, bouncing off Nero's tomb. The crown fell from the balloon and came to rest at a rakish angle on the tomb of the fiddling Roman.

This was the opening the Italian press was looking for, and they promptly took the dictatorial little Corsican over the coals. Napoleon fired his air service chief over the embarrassment. He turned to the person closest to him, Madame Blanchard, and quickly appointed her to the post. She was to be, as many historians agree, no more than a figurehead to feed Napoleon's exhibitionism. Like her sister aeronauts, however, she attracted much attention to her aerial exploits and they were a constant source of news for the French people. This attention clouded and left unnoticed what may have been her most important contribution to the fledgling science of aeronautics.

Napoleon had planned on invading England, and part of his plan was to launch a full scale troop invasion in balloons, across the English Channel. Madame Blanchard had known how easy it was for her husband to cross from England. And she knew that Phillip Rozier had the dubious distinction of being the first person to die in an aerial accident when he attempted a balloon flight across the English Channel from France, against the headwinds. She cautioned Napoleon that the invasion from the air would not work. She said if the generals doubted her word because she was a woman, they should first try it themselves. Napoleon took her advice and the generals backed down.

In 1819, four years after Napoleon met his Waterloo, Madame Blanchard was still the most popular woman in France. On the evening of July 7, 1819, Blanchard was performing a balloon-fireworks display over Tivoli Gardens. Something went wrong and the balloon went up in flames. As the balloon fell to earth, Madame Blanchard fell out and onto a roof. Her clothes were ablaze and she either fell or jumped from the roof, to her death in the street below. Blanchard's demise did not deter French women from flying and by 1834, twenty-two women had piloted their own balloons.

It took another thirty-six years before the first American woman ventured into the air. The *New York Evening Post* carried a brief report that "Mrs. Johnson" made a balloon ascent from New York's Castle Garden on October 18, 1825. She sailed across the East River and landed in the marshy flatlands in what was then the city of Brooklyn. The newspaper said she landed, "without the least injury except getting wet." There seems to be no further mention of Mrs. Johnson. The next recorded event of a woman going aloft occurred in 1855, when Miss Lucretia Bradley flew in a balloon from Easton, Pennsylvania to Phillipsburg, New Jersey.

Mary Myers

The first American woman credited with piloting her own balloon was Mary H. Myers, of Mohawk, New York. See Fig. 1-1. This flight took place on July 4, 1880 at Little Falls, New York. She billed herself as "Carlotta, the Lady Astronaut." She became famous and set an altitude record of 20,000 feet in a balloon filled with natural gas. This achievement was notable, but Mary also ascended to more than 20,000 feet without the benefit of supplemental oxygen. Mary Myers was also the co-patentee of a "Gliding Apparatus for Balloons." The apparatus, Myers claimed, allowed her to guide her hydrogen-filled balloon at will.

The newspapers marveled at her bravery and noted that, "She stepped into the basket with no more concern that most ladies would exhibit on entering a carriage for a drive"; the *Daily Saratogan* reported; "Mlle. Carlotta may be said to be a veritable high flyer."

The high-flying Mary Myers was, according to local-historians, the "handsome and highly intellectual Mary Breed Hawley Myers, a descendent of the Bostonian Breeds of Breeds Hill fame." Carl Myers, her husband and col-

Fig. 1-1 Mary Myers

laborator in many of her ventures, was a mechanically gifted aeronautics pioneer who owned and operated a balloon farm in New York's Mohawk Valley. "Carlotta" and her husband Carl conducted experiments, manufactured balloons and featured parachute drops, captive balloon rides and gave a few aerial weddings.

Mary Myers hosted "lawn parties" where thousands came to watch her land her balloon with the precision of "a foot passenger selecting his particular street car." Myers saw nothing unusual in her feats and attributed her success to her knowledge of balloon construction, the steering device she co-patented with her husband and "an abundance of health such as comparatively few women are blessed with."[2]

Common throughout the rise of women in aviation was the usual physical description of the pilot by her supporters and likewise her detractors. It was no different in 1885. One reporter described Mary Myers as, "Pretty as a picture, not at all masculine, a strong healthy lady whose features betoken great courage and decision of character." Another, perhaps not so obviously smitten, wrote of her "simple flannel dress of blue, with plain waist and no wrap... her hair neatly arranged in a double knot now so much affected, encircled by a blue ribbon gave her a plain appearance." The reporter also noted that her short-skirted and braid-trimmed blue suit revealed "neat-fitting gaiters."[3]

Between 1880 and 1890 Mary Myers completed more balloon ascents (typically lasting several hours) than any other living person and more than all the other women in the world combined.

Aida de Acosta

The fact that a woman preceded the Wright brothers into the air, in a powered machine was almost lost to history. On July 9, 1903, five months before the Wright brothers made their historic flight, the beautiful Cuban-born American, Aida de Acosta, soloed in a dirigible just outside Paris, France. The craft's owner, the famous Brazilian dirigible designer, Alberto Santos-Dumont, and others involved at the time, wanted the incident to quickly slip from prominence, and be forgotten.

Santos-Dumont had a great deal of confidence in his young student. De Acosta had learned quickly, but Santos-Dumont did not have complete confidence in her. Before her flight, he tied a cord to the escape valve, and around her waist. "If you go too high in the air and get frightened," he said, "fall to the floor of the basket. A string will pull the escape valve and the balloon will lose its gas, and come down." Santos-Dumont also had to adjust the craft to accommodate his new pilot. De Acosta's weight included a Victorian assortment of petticoats, and bustles, and she weighed slightly more than the lithe Santos-Dumont, so it was necessary to remove some sandbag ballast, and readjust the craft's balance.

As Aida sailed up into the clear, benevolent French air, she marveled at the new view her station offered. She watched Santos-Dumont as he peddled a bicycle below her. She following the hand signals from her instructor, and headed for the polo grounds. De Acosta and Santos-Dumont did not publicize the flight and when she landed in Bagatelle, she was in the middle of the country's largest polo game between England and America. People expected to see a male pilot alight from the craft. Santos-Dumont, beaming, approached the dirigible's basket and helped its single passenger out. He said, "Mademoiselle, vous entes la premiere aero-chauffeuse du monde!" De Acosta's untraditional arrival created an uproar that temporarily stopped the game.

After the crowd had settled down, officials escorted the "drop-in" celebrity and her mentor to the grandstand where they watched the rest of the game. Afterward, de Acosta attempted to return the way she had arrived, in the dirigible. The crowd objected and her friends predicted grave consequences if she were foolish enough to try a repeat performance. Undaunted, she climbed into the machine and after Santos-Dumont untied the mooring lines she sailed off into the Paris sky, landing safely several miles away.

The press did not look kindly on de Acosta as the first woman in the world to fly a powered machine and severely criticized Santos-Dumont. Acosta's parents, who believed a woman's name should only appear in newsprint twice, at her marriage and her death, were sickened by the offensive publicity. They made Santos-Dumont promise never to reveal the identity of the pilot. A polite person to the end, he was regretfully true to his promise. However, in his book, *Dans l'Air*, he describes the event. Santos-Dumont mentions that the story must be recorded in the history of flight and identifies de Acosta only as, "the heroine, a young and very pretty Cuban, prominent in New York Society."[4]

The history-making trip of Miss de Acosta faded like a sunset into the dark recesses of history. This significant event in aviation history was not re-discovered until thirty years later, by Helen S. Waterhouse, the aviation editor for the *Akron Beacon Journal*. Her full account of the first woman to pilot a powered vehicle in flight appeared in the July 1933 issue of *Sportsman Pilot*.

After the Wright brothers first powered flight in a heavier-than-air machine, the French rose to the leadership role in world aviation. In 1910, Raymonde de la Roche, of France, became the first woman in the world to earn a pilot's license. In the United States, however, the prominent men in aviation, like the Wrights, refused to let women fly. Katharine Wright, a staunch supporter of her brothers never took the controls of a flying machine. They did accept her financing of their experiments and allowed her to fly as a passenger to help demonstrate their machine to the skeptical public.

Harriet Quimby

Although Blanche Scott was the first American woman to solo in a heavier-than-air machine in 1910, the first American woman to earn a pilot's license was Harriet Quimby, in 1911. Harriet Quimby was a popular figure in early aviation. Her death in an unstable aircraft eleven months later did not make acceptance of women in aviation in the second decade of the twentieth century, any easier.

The few historical records of her life are filled with a series of contradictions and controversies. William Quimby married Ursula Cook, in Coldwater, Michigan, on October 9, 1859. William was 30 years old in 1864, when he volunteered for the Union Army. He worked as the regimental cook with the 188th Infantry from New York. After the war he and Ursula began their family. Their first daughter, Kittie, was born in 1865, and he was 41 when his daughter Harriet was born on May 1, 1875, in Ovid township, Michigan.[5]

The Quimbys moved to California around 1884, and throughout Harriet's formative years her mother's influence molded her character. Harriet spent many hours at her mother's elbow learning the ways of late 19th century California, and Ursula Quimby greatly affected her daughter's adult years. Her mother was a firm believer in women's emancipation. The years of 16-hour work days on the farm had taken a physical toll on Ursula Quimby. Her husband had worked sporadically and generally unsuccessfully at various jobs, and it was after they moved to California that Ursula Quimby took control of her family's destiny. Harriet's uncle, her mother's brother, was renowned for herbal medicine cures. Harriet and her mother mixed his formulas and sat home bottling the curatives. Her father became an itinerant salesman and sold them from a wagon. Ursula Quimby and her daughters augmented the family's income by making prune sacks for the local fruit packing industry. See Fig. 1-2.

To prepare young Harriet for the male-dominated world her mother began an image-building campaign. She told people Harriet was born in 1884, making her nine years younger than she really was. She also created a fantasy life by telling people Harriet was born in Boston, and educated in Switzerland and France.

Harriet was 25 years old in 1900, and shows up with her mother and father in the 1900 census in San Francisco. By this time, her sister Kittie was gone from the home and lost to history. By the age of 26, Harriet was working on the San Francisco *Call and Chronicle* in the copy room. Harriet would search for stories around the city, write them, and hand them to the editor. Soon the editor, Will Irwin, realized Harriet was a natural writer and remarked she had the "best nose for news" he had ever seen. He began to give her regular assignments as a cub reporter. Quimby quickly began to build a reputation as one of the paper's best reporters, and soon her byline was well-known in northern California. She was one of the first women to work as a journalist for a major newspaper.

This was a time in American society when a young woman of upper class and education looked forward to an early marriage and family responsibilities. Harriet, however, did not shock her mother when she announced plans to move east and make her mark as a journalist. Her career consumed her energy and Harriet did not let any serious or permanent romantic attachments interfere with her goals. It was difficult enough convincing narrow-minded editors to take her seriously; any hint of marriage would have immediately ended her career hopes.

The editors of the New York papers were a tough bunch, looking only for results, but they were also skeptical because she was a woman. There were not many female journalists. They were mildly interested in what she had written and published in the past, but more interested in her potential to make money for their newspapers. The editor of *Leslie's Weekly* first talked to her in terms of trying her on a few assignments and then had second thoughts. He reread one of her articles and declared out loud that she was not a literary genius. He then suggested she might consider another line of work. Harriet smiled warmly at him when he had finished his verbal assault and replied that that was precisely the same opinion she had of herself. However, she had tried cooking and she could write better than she could cook, she told him. She then went on to explain that the two of them would get along fine. The editor recognized her quick wit and assertive posture and he was so taken back by her reply, he hired her on the spot, as a part-time writer. She had the makings of a good journalist. It wasn't long before Harriet was a full-time reporter.

In the early 20th century, the horse and carriage were marks of elegance, but the automobile was creating a new social class. One day, Harriet went to an auto race on Long Island and convinced a race driver to take her around the track in his racing car. Her heart pounded, and the 60 mph speed in the open

Fig. 1-2 Harriet Quimby

car nearly took her breath away. This was her first taste of danger associated with speed and Quimby found it intoxicating. One racing car experience and the catharsis of writing about it did not cure Harriet of the fever for adventure.

Harriet Quimby's gregarious personality enabled her to be friends with people from all walks of life. One group she befriended was a small but ardent gaggle of fledgling aviators. In the fall of 1910, Miss Quimby, now 35, had gone to Belmont Park, on Long Island at the invitation of her closest friend, Matilde Moisant. Quimby was eager to travel anywhere for a story, and was always looking for something dramatic to share with her readers. The Belmont Park Air Meet, the occasion for an air show, was her first encounter with an airplane. Belmont Park had evolved with early aviation into the 1910 "Jet Set" and everyone who was anyone in society and aviation attended. Some spectators saw the future of America and their future linked to these fragile wooden vehicles. Others saw the event as a social gathering. A few went expecting to see a fiery crash. Quimby watched in amazement as the "birdmen-heroes" as she later called them - flew loops and dives and posed for pictures in their biplanes and monoplanes. The event attracted the world's greatest aviators and included the elegant Englishman, Claude Grahame-White, the Frenchman Herbert Latham, and the Americans, Glenn Curtiss, Ralph Johnstone and Arch Hoxsey.

After seeing these odd-shaped contraptions in the sky, Miss Quimby concocted an angle for a story. It was October 30, the day before Halloween -

and Quimby could link these odd-shaped machines roaring through the skies with the more traditional gremlins, ghosts, and goblins.

Matilde told Quimby that her brother John was in an event - the air race. Now that sounded exciting to Harriet! It *was* exciting, and Quimby was not disappointed. Harriet witnessed John Moisant, a wealthy investor, fly a monoplane from Belmont Park to the Statue of Liberty and back. He flew the 36 miles in 34 minutes, winning the Statue of Liberty Race, and defeating the best pilots Europe had to offer. The air show convinced Harriet that flying was more fun than driving. "It really looks quite easy," she told her friend, Matilde. "I believe I could do it myself, and I will."

In the spring Quimby's passion for this new form of adventure had not faded, although a tragic event had taken place. In early January 1911, Harriet and her friend Matilde received word that Matilde's brother John died in an airplane accident near New Orleans, on December 31, 1910. That may have ended plans to take flying lessons, but not for Harriet Quimby or Matilde Moisant! Although shaken by the death of his brother, Matilde's brother Alfred decided not to close their school. Americans involved in aviation recognized that they were behind European aviation. Alfred Moisant said he wanted to, ". . . raise aviation in the United States from its condition of stagnation or worse to the place it occupies abroad."[6] On May 10, 1911 Harriet and Matilde enrolled in Moisant's aviation school on Nassau Boulevard, on Long Island. They had their first lesson the same day.

Due to the social pressures toward women in that period, the two women dressed disguised as men for their lessons. Harriet always took her lessons at sunrise when the lessons didn't interfere with her work. The air was usually calm at this time, and she could keep her activities a secret - or so she thought. When a reporter discovered Harriet's charade, she was happy to see it end. The newspapers gave her much publicity. The press described Harriet as a "willowy brunette" and they quickly tagged her with the nickname the "Dresden China Aviatrix," because of her "beauty, daintiness, and haunting blue eyes." Quimby quickly became well-known and garnered publicity from the newspapers. "She was a strong woman," said Matilde Moisant. "Harriet enjoyed the publicity."

On July 31, 1911, Harriet was ready for her pilot's exam. In her article, "How I Earned My Pilot's License," Quimby revealed that she had 33 lessons and the actual time spent in each lesson was between two and five minutes. She had spent a total of less than two hours in the air. She pointed out that was the allotted time for students at each lesson taught in the leading French aeronautical schools. Her course of instruction covered many weeks due to adverse weather conditions. There were many days when the wind prevented even the most experienced pilots from going aloft. In the article, she attributed her success to a competent instructor, and the "kindly" hand that fate dealt her.[7]

Quimby described in another article what she thought were prerequisites for the prospective student pilot. "One who has run a motorcycle or an automobile successfully is all the better qualified to begin his lessons as an aviator. Without experience of this kind, the noise of the unmuffled motor in an aeroplane will be nerve-wracking."[8]

Harriet Quimby completed two thirds of the pilot's exam with ease. The first two parts consisted of completing five alternate right and left turns around pylons, and completing five figure eights. The third part was a landing test. It required her to land within 100 feet of where the plane had left the ground. On her first attempt, she landed too far from the spot. It was too late for another attempt. Tomorrow Harriet would try again.

Dawn

Just after dawn, On August 1, 1911, Quimby climbed into the single seat craft. There was a small crowd gathering, attracted from nearby Mineola and Garden City, by the fast-spreading rumor that a woman was attempting to earn a pilot's license. There was silence as the mechanic primed the engine by turning or "pulling through" the propeller. Two men held the rudder, and another two held the landing gear. Harriet checked out her "controls" - only a warping lever/control stick and rudder bars. The primitive craft lacked ailerons, so to raise and lower the wings Quimby had to twist the lever that, in turn, warped the entire fragile wood and silk-covered wing. Quimby flipped a switch and raised her arm - the signal for the mechanic to give a strong swing to the wooden propeller. The 30 horsepower Anzani engine coughed to life, its propeller spinning blindly at 1,400 revolutions per minute (rpm). Its three cylinders spewed black oil-filled smoke on the four men straining to hold back the aircraft. When the craft "felt right" Quimby raised her hand in a thumbs-up position and the men released the plane. The plane catapulted forward, bouncing down the rut and bump-filled grass strip for about 50 feet. On one bounce, it leaped upward. Quimby pulled back on the stick that moved two cables which in turn operated the elevating surface on the tail. The craft was flying!

Quimby was at an altitude of about 150 feet, and all was going well. Her ground speed was about 45 miles per hour when she executed the first of her turns. Turning in a Bleriot-styled monoplane required a high degree of dexterity and coordination. Quimby had to keep her eyes on the horizon, and on the wing tip to keep the plane level. She also had to maintain the wings in an absolute level attitude while turning the craft. This required making a flat skidding turn, using the rudder. The sound of the engine was the only indicator that the three cylinders were firing properly.

It was time for the landing test. The previous day, she had landed and, in her excitement, shut down the engine too soon. (There were no foot brakes in those days.) This time Quimby planned her landing from the air. She mentally marked her touchdown spot, took aim at the white square, and began to increase her glide slope. The plane's wheels touched down perfectly. She counted off three seconds then flipped a switch that shorted out the spark plugs and shut down the engine. A silence suddenly fell over the area. All Harriet could hear was the monoplane's creaking spruce framework as it bounced along the grass strip. The craft gradually slowed to a stop - just seven feet, nine inches from the mark. Quimby had set a record! And the crowd cheered her accomplishment.

"After the flight," Quimby said, "I removed my goggles, then climbed out of my monoplane. The crowd was applauding and it sounded good to me. I waved to everybody and nonchalantly walked over to the official observer. I looked him in the eye and said, well I guess I get my license."

"I guess you do," was his reluctant reply. On August 1, 1911, Harriet Quimby became the first American woman to earn her pilot's license (#37).[9]

Matilde Moisant received her license 12 days later, setting a record for students that remains unbroken: Moisant spent a total of 32 minutes in the air before qualifying for her pilot's license! [10]

Years later Matilde confided to Bobbi Trout, another pioneer aviator, that she let Harriet try first for her license since Quimby had declared her intention to earn a living in aviation.[11]

Harriet Quimby had many vocal critics concerned about her "immodest dress." Once Quimby began to appear in public, her original flying outfit pieced together from men's clothing was simply unacceptable to her. Her critics said she was corrupting the public's morals wearing men's clothes. Quimby was not happy with the idea either.

Quimby and other women who dared to fly received definite signals from the establishment on what was inappropriate dress, and what roles women were to assume. Even her own newspaper took a position. *Leslie's Weekly* ran an editorial on October 5, 1911, commenting on a sermon delivered by a Bishop Nilan. "The bishop was particularly critical, and justly so, of the vulgarity of the costumes worn by women today. The fashion designers are preparing women to take her place with man by shaping her garments so that they often closely resemble man's attire. Wearing these she disfigures her beauty and deforms her nature. Gone is the old-time womanly expression of sweetness and modesty, and in its place we have its swagger and stare. In the prevailing style of dress women, without question is exposing her figure as never before in the history of civilization. Had the change come about all at once, it would have shocked the sex into a revulsion of feeling; but coming gradually, it has taken advantage of the weakness of women to accept the style as handed down with too little protest. Through a little independence in their dress and that of their daughters, mothers of refinement may yet save the day for modesty and even virtue." Of course the editor missed the point. Some women were exercising their independence by wearing what they pleased.

Quimby approached her critics and this challenge with an air of dignity, flamboyance, and style. "It may seem remarkable," she said, "but when I began to fly I could not find a regular aviator's suit of any description in the great city of New York, and I tried hard. In my perplexity, it occurred to me that the president of the American Tailor's Association, Alexander Green, might be a good advisor; he was. It did not take him long to design a suit that no doubt established the aviation costume for women in this country, if not the world, since the French women continue to wear the clumsy and uncomfortable harem skirt as a flying costume." The outfit was extraordinary for 1911; a one-piece purple satin outfit with full knickers reaching below the knee, and high laced black kid boots. Her head gear resembled a monk's hood, and her accessories were flying goggles, elbow-length matching gauntlet-style driving gloves, and a long leather coat for cold weather flying. In colder weather, she wore a full length cape to match the purple outfit. The hood had small swatches of black net inserted over the parts that covered her ears, so she could "keep in touch" with the workings of the engine. "It was also an ingenious combination," she said. "It can be almost immediately converted into a conventional-appearing walking skirt when not used in the Knickerbocker form."[12]

Soon after earning her license, Quimby earned her first professional fee: $1,500. For most pilots in those days, earning money was difficult. A license was needed to appear at professional air meets, and unless the pilot owned an aircraft, it was often a hand-to-mouth experience. A pilot could be expected to negotiate between 20 and 35 percent of the net earnings of the meet. However, the pilot had to pay for mechanics, transportation of the craft (on a train) to each meet, hangar fees, fuel and tips to all the tag-alongs who followed waiting to minister in an awkward form of hero worship. A pilot felt he was ahead of the game if his net earnings were 10 percent. The lion's share of the profits always went to the manufacturer or suppliers of the machines.

Harriet set her first official record at her first air meet. On September 4, 1911, with a crowd of 20,000 persons gathered at the Richmond County Fair, (at Dongan Hills, on Staten Island), she became the first woman to make a night flight.[13]

Flying the English Channel

Harriet Quimby wanted to show women they could branch out and try new experiences. Quimby decided that she would be the first woman to fly across the English Channel.

When Quimby arrived in England with her business manager, A. Leo Stevens, they approached the London *Daily Mirror.* The *Mirror* enthusiastically agreed to be a co-sponsor and back her efforts with, as Quimby put it, "a handsome inducement" for exclusive European publication rights to her flight. (*Leslie's Weekly* was her other sponsor.)

Quimby and Stevens then sailed across the Channel to France to meet Louis Bleriot, in Calais, to negotiate an aircraft for the flight. "I never was the best kind of sailor, and there was a real satisfaction in contemplating the crossing in the air and mocking the waves that had so often made me uncomfortable."[14]

Quimby arranged for the airplane to be shipped in secret to an airfield in Dover, on the coast of England. No sooner had she registered when she discovered that another woman had flown across the Channel. Just days before, Eleanor Trehawke-Davis had hired Gustav Hamel as a pilot and had flown as a passenger across the Channel. It bothered Quimby that someone on the *Mirror* must have leaked the story, but she was more annoyed that they had also withdrawn their exclusive coverage and "handsome inducement" ($5,000). This setback, Harriet decided, would not stop her. She could still be the first woman to pilot a plane across the Channel. Quimby was too focused on her goals to hold a grudge. Men had been making the trip for almost three years, ever since Louis Bleriot had made the first trip on July 25, 1909. While there were women flying in Europe at the time, none had dared to challenge the Channel. Quimby believed being the first woman to pilot a plane across the Channel was a bigger prize, and the prize was hers if she acted quickly.

On Sunday, April 14, Quimby and her party drove to the aerodrome three miles outside Dover to look at the newly assembled monoplane. She wondered about its control. Harriet had learned to fly in a 30 horsepower, Moisant-built plane, but this plane was genuine 50 horsepower Bleriot, and Louis Bleriot and Gustav Hamel said it would handle differently.

The weather was perfect. The air was clear, and she could just barely make out Calais, 22 miles across the Channel. Those present - including the reporters - urged her to make a practice flight and take off immediately for Calais to take advantage of the weather. The weather forecast for Monday was for high winds and possibly rain. Previous experience with flights from the Dover aerodrome had shown that absolutely still air was necessary. To their dismay, Harriet refused to fly. It was Sunday and she had promised her mother she would not fly on that one day for any reason.

Monday brought thick clouds, and heavy rain over Dover and the English Channel. "I was not nervous about what lay ahead of me. I was impatient to get going, despite the protests of my friends," Quimby recalled. "I was going to fly on the other side of the Atlantic for the first time. My anxiety was not getting started quickly."[15] Harriet and her small ground crew sat all day in a damp hangar at the aerodrome, waiting for the weather to clear. The reporters also waited, but slowly they began to leave, convinced not even a lunatic would fly in wind and rain.

Hamel had successfully completed the trip three times and knew the dangers of such a trip. He knew too that there was a vast difference between crossing the Channel as a passenger with an experienced pilot and making the crossing alone for the first time. The evening before the trip, Hamel tried to dissuade Quimby. "He told me that a woman could not make the trip alone." Hamel was anxious for Quimby's safety and questioned her ability to pilot the airplane across the Channel. "Hamel suggested dressing up in my flying costume, flying across the Channel and landing at a remote spot where I would be waiting to take the credit." Quimby resolutely refused Hamel's offer. Hamel then offered to show Quimby how to read a compass, an offer she did accept. A compass was something new to her. "I was annoyed from the start by the attitude of doubt by the spectators that I would never really make the flight. They knew I had never used the machine before, and probably thought I would find some excuse at the last moment to back out of the flight. This attitude made me more determined than ever to succeed."[16]

Takeoff

It was 3:30 Tuesday morning, April 16, 1912. Quimby arose and had a cup of strong British tea and biscuits for breakfast. Her small party then got into automobiles and by five o'clock were on the flying grounds. The rain had stopped, but there was patchy fog in the area. The early morning air was chilly, but thankfully there was no wind. The ground crew pushed the monoplane out of the hanger. Swiftness was important, for it was almost certain that the wind would rise again, within an hour. Hamel volunteered to test fly the new machine. He was still feeling some guilt over being an accomplice on Mrs. Davis' trip. He jumped into the machine and was off for a short tryout of the engine and to report the atmospheric conditions. The machine checked out perfectly. "Now it was my turn," said Quimby. The sky immediately over the aerodrome was clear, but the French coast, unlike two days ago, was obscured by a moving wall of mist.

Quimby took off at 5:30 a.m. aiming her plane for France. "I could see people waving but the engine roaring at 1200 revolutions per minute drowned out what surely were cheers and wishes for my safe journey."[17]

Later Quimby reflected on her trip. "On takeoff, I saw at once that I had only to rise in my machine, fix my eyes upon Dover Castle, fly over it and speed directly across to the French coast. It seemed so easy, like a cross-country flight. I am glad I thought so, otherwise I might have had more hesitation knowing that the treacherous North Sea stood ready to receive me if I drifted off my course." D. Leslie Allen, an English aviator, started the same day as Quimby, in a monoplane similar to the Bleriot, on a flight over the Irish Channel, from London to Dublin. [18] He was never seen again. That was the mystery and danger of an over water flight in those days.

Hamel had only given Quimby brief training in reading a compass. She had never seen one operating in a moving plane. Hamel told her it would be shaking from the engine vibration, and difficult to read. "I was hardly out of sight of the cheering crowd before I hit a heavy fog bank, and found the compass to be of invaluable help." Quimby had only two instruments with her: She wore a watch on her wrist, and held the compass between her knees. Quimby could not see above, below, or ahead. "I started climbing to gain altitude, hoping to escape the fog," she said. It was bitter cold, the kind that chills to the bone. I climbed to a height of 6,000 feet looking for a hole to escape the mist that engulfed me." Under her flying suit of wool-backed satin, Quimby wore two pairs of silk combinations. Over her flying suit she wore a long wool coat, and over that, an American raincoat. Around her shoulders she wore a long wide stole of sealskin. Even this did not satisfy her solicitous friends. At the last minute, they handed her a large hot water bag, which Hamel tied to her waist.

Quimby also wore Scotch woolen gloves that gave her good protection from the cold, but the machine was wet and her face was covered with dampness. "I had to push my goggles to my forehead since I could not see through them. I was traveling at a mile a minute and the mist felt like tiny needles on my skin," she said. [19]

During the flight, Harriet recalled Hamel's remark about the North Sea. If she drifted off course by as little as five miles, she would get lost and probably go down in the icy waters. (Hamel should have taken his own advice as seriously as he had given it to Quimby. Later he flew off into the Channel mist and never returned.)

Victory

Around the mid-Channel point, the fog was beginning to take its toll on her nerves. Her head ached from the strain of keeping the craft level. Her senses were keenly aware of the attitude of the plane. The concentration on the task was exhausting. The minutes ticked by like hours as she listened to the sound of the engine. If the engine should quit, she knew she had little chance of surviving a crash-landing in the cold, fog-shrouded waters of the Channel. She decided to descend and look for clear air. As she lost altitude, the machine tilted to a steep angle, causing the gasoline to flood the engine. (This was a design flaw that had proven fatal for several pilots.) The machine began to backfire. Quimby regained control of the craft and began to consider her options. She had but one choice, to land on the water. (The plane had a balloon-like floatation device running the length of the skeletal frame. It is doubtful the airplane would have floated very long.) "To my great relief," she said, "the gasoline quickly burned away and my engine began an even purr. I glanced at my watch and estimated I should be close to the French coast."[20] Soon the little Bleriot broke through the mist and into the brilliant sunlight. Gleaming white sand flanked by green grass caught Quimby's eye. The fog had disappeared and she knew she was safe. Because of winds and the underpowered 50 horsepower engine, the 22-mile trip took one hour and nine minutes.

Harriet recalled seeing landfall. "I flew a short distance inland to locate myself or find a good place on which to land the machine. I was happy to be ashore but I could not find Calais. Meanwhile the wind had risen and the currents were coming in billowy gusts. The land below me was all tilled and rather than tear up the farmer's fields I decided to drop to the hard sandy beach. I did so at once, making an easy landing." See Fig. 1-3. [21]

Eight decades later it may be difficult to understand the danger in the 22-mile Channel flight. Today we can put Harriet Quimby's accomplishment in proper perspective. Only nine years had gone by since the Wright brothers' first powered flight in a heavier-than-air machine. Bleriot's monoplane had a reputation as the trickiest aircraft the French designer had yet produced. Several flyers attempting the flight had been lost over the Channel, and by the 1912 rule of thumb of not flying in fog, rain, or clouds, at night or in more than a five mile per hour wind, Quimby should have turned back. There were no parachutes in those days, no guidance equipment, radios or navigating charts. A flight over

Fig. 1-3 Quimby seen here after her successful flight

water, out of sight of land, was perilous. For Quimby this was her first over water flight, fog and rain obscured most of her trip, and it was her first use of a compass, in an unfamiliar plane. In addition, she flew a craft that warped its wings, and its engine needed an equal amount of gasoline, prayer, and luck. Her airplane had no instruments, and was scarcely more than a winged skeleton with an under-powered engine. She had to keep the Bleriot on an even keel, with no horizon as a frame of reference. Surely, for anyone to fly across the 22-mile English Channel in 1912 required extraordinary courage, skill, and self-confidence. Quimby, however, minimized her feat. "The trip was as easy as sitting at home in an arm chair," she said, "and I never had any doubt of my success." On another occasion she said, "Any woman with sufficient self confidence and a cool head could fly across the Channel as easily as I did. Within months, probably weeks some other woman probably will make the same flight or even achieve some greater undertaking."[22]

The Boston Air Meet

Quimby had been working toward her goal of financial independence through the growing popularity of air meets. The Boston Air Meet, on Squantum Bay, attracted the biggest names in aviation. Other women, like Blanche Scott and Ruth Law, had registered in the event, but Quimby was the main attraction.

The event featured many top aviators; Lincoln Beachey, Glenn Martin, and Earl Ovington were among the top names. The promoters of the air meet had money on their minds. So did Quimby's manager. He drove a hard bargain. Stevens had a unique package to sell. Quimby was the world's best known female aviator, it was her first public appearance since her victorious Channel crossing, and she was flying a new and more powerful Bleriot monoplane. Her trusted friend, Matilde Moisant, described another reason Stevens was a tough negotiator. "Harriet was the prettiest girl I have ever seen. She had the most beautiful blue eyes - oh what eyes she had. She was a tall and willowy brunette and when she wore her long cape over her satin plum-colored flying suit, she was a real head-turner." The meet's manager, William Willard, agreed that Quimby was an irresistible drawing card, and reluctantly agreed to Stevens' terms, ($100,000) for Quimby's performance. The other top-billed male aviators also tried unsuccessfully to negotiate what were outrageous sums in those days.

Quimby was proud of her flying record. The fragile airplanes of the day would sometimes break up in flight and failure of the unreliable engines had caused many accidents and fatalities. The Bleriot had a wing-warping device that in the hands of a pilot who overreacted could literally twist the wings off the plane. Quimby never had a flying accident, because she was a careful and capable pilot. She gave close attention to every wire and fastener before each flight. See Fig. 1-4.

Quimby's new 70 horsepower two-passenger machine, originally designed for military use was fast, could climb easily - but it was also difficult to control. The balance of the airplane was critical. With no passenger in the rear seat, a sand bag provided artificial ballast. The gyroscopic effect of the big Gnome engine with cylinders revolving around the drive shaft was also known to cause control problems. It took all of Harriet's skill and experience to successfully fly this Gallic beast.

Tragedy

William P. Willard, the manager of the event, had won the flip of a coin and the honor of flying with Quimby as the last show of the day. Willard headed for the Bleriot, eager to get airborne. Quimby emerged from the hangar dressed in her famous plum-colored outfit, fingering her lucky jewelry, as if to waken the good spirit within the amulet. Immediately, reporters rushed to her, looking for an interview. Quimby obliged and ended the plane-side conference by assuring reporters that a forced landing was out of the question. "I have no intention of coming down in the water," she said. "I'm a cat, and I don't like the water." With that she climbed the portable steps to the cockpit, and with a dramatic flair, tossed her cape to Stevens, and stepped into the pilot's seat.

With Willard seated comfortably in the back, Quimby gave the signal and her mechanic pulled the propeller through one turn. Quimby flipped a switch, and the mechanic spun the propeller, again. The engine choked and sputtered as the cylinders fired. The propeller whirled to life. The plane was shaking as the powerful rotary engine tried to pull the plane forward. Four men held the machine as it strained forward. Quimby pushed the throttle, increasing the gas

Fig. 1-4 Harriet Quimby at the Boston Air Meet

flow to the engine. When the plane felt right, Quimby gave the signal to the men. They let go of the airplane and Harriet was into her takeoff roll. Moments later, the Bleriot lifted gracefully into the air and Quimby headed for the 27-mile course around the Boston Light.

The outbound flight was uneventful and ten minutes later Quimby flew past the Boston Light. With a gentle touch on the rudder, and a slight warp of the wings, she came out of a descending turn around the lighthouse at an altitude of 2,000 feet. The sun was beginning to set, turning the sky brilliant orange. Everyone watched the silhouette of the dragonfly-like monoplane against the fiery sky. Suddenly the plane's tail rose sharply, tossing Willard out of the craft and into an arc that quickly turned to a vertical fall. Observers reported Quimby fought to regain control of the craft. For a moment the monoplane seemed to slide back towards a normal attitude, a split second later, the monoplane's tail reared up again, standing the plane on its nose. As the plane went perpendicular,

Quimby's body catapulted from her seat. The plane continued rolling over on its back and fell toward the bay.

Thoughts of beauty disappeared as five thousand horrified people watched as Willard and Quimby tumbled through the air and plunged into the harbor waters 200 feet from shore. The tide was low and at the time the water was only four feet deep where they fell. Quimby died on impact, and Willard drowned. Ironically, the Bleriot monoplane flew itself out of the dive, and glided into the water. As the plane's wheels touched the water, it tripped on its landing gear nosed over on impact, but sustained little damage.[23]

The Technical Explanation

In August 1912, *Aircraft* magazine devoted four pages to the accident. One article convincingly argued the dangerous instability of that monoplane design. The author pointed out that the fixed horizontal tail surface of the plane was actually a small cambered wing set at a higher lifting angle very similar to the craft's main lifting wing. "A machine of this type," wrote the author, "has not the slightest degree of automatic longitudinal stability. It is an extremely tricky and dangerous type to handle. The horizontal tail should act as a stabilizing damper, preventing the machine from either diving too steeply or stalling. Not under any circumstances should it act as a lifting plane." The article listed almost a dozen pilots who died in Bleriot monoplanes under similar circumstances where the plane dove straight into the ground. (He does not say if any of the victims fell from the craft as did Quimby and Willard.)[24]

With today's knowledge of aeronautics, the 1912 analysis is accurate. The Bleriot monoplane was a terribly unforgiving aircraft, even for pilots like Quimby who had more experience in the type. In hindsight, Harriet Quimby and her unsuspecting passenger were truly victims of aviation's age of innocence.

Alys McKey Bryant

Few women braved the male-world of aviation in 1912. But those that did, were daring. Within 60 days of Harriet Quimby's death, Alys McKey, an energetic and practical woman, picked up where Harriet Quimby left off. McKey drove an automobile, rode a motorcycle, and watched with sharp brown eyes for any challenge that offered something different. She answered such an opportunity in a classified ad placed in a California newspaper. "Wanted - a young lady to learn exhibition flying." After Johnny Bryant, the chief pilot for the Bennett Aero Company, interviewed McKey, he signed her on as a pilot to do exhibition flying with his team. It didn't matter that McKey had never been near an airplane. In those days, many pilots had to teach themselves to fly, and she was no exception. McKey, however, did get tougher breaks than she deserved. When Bryant introduced her to the plane she would fly, she noticed it was "rolled up in a ball" as pilots referred to a wrecked airplane. Bryant had crashed the day before and the craft needed complete rebuilding. Only after McKey had rebuilt the machine herself, did Bryant allow her to fly the craft. On one of her "grass

Fig. 1-5 Alys McKey Bryant

cutting" flights, where she skimmed along the ground just above the grass, a wheel touched down in a gopher hole, and the plane ground-looped. Again McKey had to rebuild the plane. This time, when the plane was ready for flight, the plane's owner, Fred Bennett decided to take the plane back to its home field. McKey and Bryant were to fly the plane with Bennett squeezed into the fuselage. On takeoff, the plane veered to the left and because of the poorly distributed weight, dove into the ground, again! See Fig. 1-5.

Undaunted, McKey went back to the lumber yard for the raw materials. She had developed a passion for anything mechanical, or labor intensive. She relished in building things and throughout most of her aviation career, automobile and motorcycle days, McKey performed the manual labor common to men in those days. In November 1912, McKey finally soloed. In accomplishing this she had fulfilled a lifelong dream to fly. When she was twelve years old McKey had written an essay for school about imaginary flight from New Jersey to California.

McKey performed her first exhibition flight on May 3, 1913, in North Yakima, Washington. She also became the first woman to fly in Washington, Oregon, Idaho, and Canada. McKey secretly married Johnny Bryant on May 29, 1913. The marriage seemed to motivate her even more to achieve her own identity in aviation. McKey promptly set a new women's altitude record of 2,900 feet. Her marriage to Johnny Bryant was happy but short-lived. She was present at an exhibition he was flying when the steering mechanism on his plane broke. She watched helplessly as he crashed to his death, just ten weeks after they had been married. [25]

With the death of her husband McKey went into retirement. But like most people in aviation, she could not stay away. She was soon back to work as a stunt pilot for the movies. [26] After that she moved east to St. Louis to put her mechanical ability to work for the Benoist Airplane Company. From there she went on to fly as a test pilot and instructor. When the government rejected her repeated application to be a pilot in the Army Air Service during World War I, she turned to the Goodyear Company in Akron, Ohio to help in the construction of balloons and dirigibles.

Her own words best describe the type of craft she and others risked their lives flying. "The wings of our plane were single-surfaced, and that one covering was of cheap, unbleached muslin. It was stretched over the top of the wing sections and fastened to the leading edge with small steel tacks and strips of split bamboo. The trailing edge, laced with piano wire, was fastened at intervals to the ends of the wing ribs. The muslin, stretched tightly by hand and tacked to each rib, had rows of split bamboo tacked and glued to make the covering more secure. The ribs and spars exposed to the weather on the underside of the plane were shellacked before the covering was fastened on, and the entire section was later treated with homemade dope. This dope left a rough, bumpy surface that would be taut on dry days, but limp in moist weather. Later, to improve the performance of our planes we put covering on the lower sides of the wings." [27]

Alys McKey Bryant never got around to getting a pilot's license and is one of only six women accepted into an exclusive club, the Early Birds. (The others were Katherine Stinson, Marjorie Stinson, Ruth Law, Blanche Scott, and Matilde Moisant.) Those women all had common traits, and one of the chief reasons for their success was their lack of fear. "Fear" said Katherine Stinson in her later years, "as I understand it, is simply due to lack of confidence or lack of knowledge - which is the same thing."

When World War I broke out, the government curtailed all civilian flying. The Stinson sisters and women like Ruth Law and Blanche Scott, petitioned the United States government to allow them to join the Air Service as pilots. There were no laws at the time to restrict them, but the government turned their request down. Hurt but not embittered by the rejection, the women went on to either fly as civilians in war bond drives or went overseas to the war zone as nurses. Marjorie Stinson took a drafting position in the Navy's department of aeronautical design.

Laura Bromwell

Laura Bromwell had her first exposure to an airplane when she saw Ruth Law at a Liberty Loan Drive. Bromwell decided that she was going to learn to fly, too. After the government lifted the ban on civilian flying she wasted no time. On October 22, 1919, Laura Bromwell became the first woman after the war to earn her Aero Club pilot's license. The next day an advertisement appeared in the *New York Times* announcing, "Women Aviators Wanted." The New York City Police Reserves were looking for women between the ages of 18 and 25 for a Women's Aviation Corps to be affiliated with the Reserves.[28] See Fig. 1-6.

The Corps was a voluntary organization under the financial sponsorship of the millionaire-advisor to the Police Department, Rodman Wanamaker. Wanamaker and others had hoped to use the corps of volunteer pilots to control traffic and transport or pursue criminals. Wanamaker's idea never got the financial support of the City of New York, and the Women's Aviation Corps soon disappeared. The men in the aviation reserves signed up with the Navy Reserve. Bromwell remained and, like the others, did not get paid for her flying, but did not have to pay for her flying either. She received a commission in the Police Reserves as a lieutenant and drew moderate attention from her unique position.

On February 17, 1920, Laura also became the first woman to fly a commercial flight over New York City. She flew a Curtiss *Oriole* and dropped leaflets in the theater district announcing the opening of the movie "Fly High! With Locklear in the Great Air Robbery."

Fig. 1-6 Laura Bromwell

Laura Bromwell may be considered one of the first women to fly an aerobatic routine. On August 20, 1920, at the rededication of the Curtiss Aerodrome at Mineola, New York, an astonished crowd watched as she looped-the-loop 87 times, breaking the record of 25 held by Adrienne Bolland of France. Bromwell counted over 100 loops but clouds partially obscured her plane from observers and they would only credit her with what they saw. On May 21, 1921, to validate her original claim, she flew 199 confirmed and consecutive loops before a crowd of 10,000. Laura landed only when she ran out of fuel.[29]

Bromwell had wanted to be the first woman to fly across the United States, but she had not flown the craft she was intending to use. Friends told her not to try aerobatics in an unfamiliar plane, but she disregarded their advice and went out of control while completing a loop. Laura Bromwell died in the crash on June 5, 1921, just two weeks after she had set her loop record. Investigation later revealed that she did not wear her seat belt. She was small and her height and the belt restricted her access to the rudder pedals.

Georgia "Tiny" Broadwick

With the glut of aircraft and pilots, and a severe recession in the fledgling aviation industry, there were few aviation jobs for women in the early 1920s. Barnstorming was the only avenue open to both men and women. State and county fairs, carnivals and motion picture companies were always looking for daring aviators, and barnstorming became very popular. Yet, flight lessons were expensive and women often took one and two jobs to afford their flight training. One job that paid enough for a woman to continue flight instructions and allowed her access to the aviation community was parachute jumping.

On June 21, 1913, Georgia "Tiny" Broadwick became the first woman to make a parachute jump from an airplane. Glenn Martin piloted the plane that traveled at a ground speed of about 60 miles per hour. At a height of about 1,000 feet, Tiny, sitting on a seat attached to the right wing, pulled a lever and the seat folded away allowing her to fall straight down. See Fig. 1-7. The static cord automatically opened the chute as she fell. She landed on her feet in the middle of Griffith Park Aviation Field, in Los Angeles. Tiny had made many parachute balloon jumps so falling through the sky was not new to her. During her earlier aerial adventures Tiny Broadwick also became the first person to demonstrate a parachute to the U.S. Army, in 1915, at San Diego, California.

Tiny also holds the record as the first person to perform a freefall in a parachute. After having problems with the automatic chute release on a previous jump, Tiny dropped from the plane and pulled the parachute release manually. Ironically this new technique went unnoticed. Five years later, when Mr. Leslie Irvine did the same thing he received national acclaim.

Tiny never got around to getting her pilot's license although she did fly some of Glenn Martin's aircraft. In 1915, she married a sea captain and after that made only occasional jumps. In 1922, she retired from aviation with more than 900 jumps to her credit. She felt that the novelty of parachute jumping had worn off.

Fig. 1-7 Tiny Broadwick

Tiny Broadwick was right. Parachute jumping was almost anticlimactic. Once the chute opened, there was no more thrill for the audience. The next level of daring that attracted public attention was wing walking.

Lillian Boyer

Lillian Boyer was looking for a career that would take her away from the boring waitress job she held, so she turned to wingwalking at air shows. Boyer had taken her first plane ride at the invitation of one of her customers. On her second ride, on April 7, 1921, she climbed out on the wing and started her career as an aerial exhibitionist. On October 10, 1921 she made her first plane-to-plane transfer and soon signed on with a barnstorming team. Her career lasted until 1929, when federal regulations on low flying forced her and other barnstormers into retirement. In the few years she barnstormed; Boyer set an enviable record of performing in 352 shows, in 41 states and Canada. In addition to her wingwalking stunts she made 143 automobile to plane changes and 37 parachute jumps.[30]

Chapter Two

The Roaring Thirties (1930-1939)

I was one of two women who was getting out there and really flying. I was the only movie stunt pilot they had.... I was the only woman around who could do that kind of work. Pancho Barnes

The gradual build-up of women in aviation during the 1920s led to hundreds of women participating in the aviation wave of the 1930s. Soon the roar of their engines was attracting nationwide attention. It was an era of intense publicity and excitement for women in aviation. The 1930s also saw the first real progress for American women in aviation. Amelia Earhart had become the spokesperson for women in aviation. Her recurring message to the people of the United States was; flying is safe, and women make good pilots. The two ideas were inseparable.

Phoebe Fairgrave Omlie

Phoebe Fairgrave was one pilot who demonstrated Amelia's ideas. Fairgrave had read about the air war in Europe and aviation intrigued her. After graduating from high school in 1920, she wanted to learn to fly; but the male flight school operators would not take her seriously. Phoebe, however, would not take no for an answer. The glut of surplus World War I aircraft gave her the answer to her dilemma. "It was not long before I began to realize that one way I could be certain to get up into the air would be to buy an airplane. I had a hunch that the prospect of a sale would make the boys waver in their determination not to have anything to do with satisfying my ambition to go aloft.[1]

Fairgrave took a small inheritance and went out and bought an airplane. She found, however, that while the "boys" would give her flight lessons, the lessons and maintenance on the machine were far more costly than she could afford. In order to raise money to keep her passion alive she took a job as a wingwalker

Fig. 2-1 Phoebe Omlie & her husband

and parachutist. She made her first jump on April 27, 1921 and by July 10, 1921 Phoebe had set a new world record for women when she jumped from 15,000 feet.

Phoebe Fairgrave further enhanced her reputation by performing variations on the basic parachute jump. After the first chute opened, she would cut it loose and freefall before deploying the second emergency chute.

By the time she was twenty, Phoebe had saved enough money to become the first woman to form a flying troupe, the Phoebe Fairgrave Flying Circus. Phoebe married the man who had taught her to fly, Vernon Omlie, in 1922. The two continued to barnstorm until they had saved enough money to open their own flying school and aviation business. Phoebe Omlie earned a living in aviation for more than twenty years and showed American women some of the possibilities. See Fig. 2-1.

Phoebe would go on to become a charter member of the Ninety-Nines, and become active in the air marking campaign of the 1930s. One of the lasting achievements of the Ninety-Nines and women like Phoebe was the successful completion of the aerial marking campaign across America. Phoebe Omlie had suggested that an air route without markings was like a highway without signs. As a member of the National Advisory Committee on Aeronautics (NACA) in 1936, Phoebe was able to get people to listen to her suggestion, and she gained the support of a very influential person, Eleanor Roosevelt. Mrs. Roosevelt influenced the Bureau of Air Commerce to hire women pilots for the air marking project. Their job was to find suitable buildings and negotiate with local officials and the owners of the buildings, to have letters painted large enough to be seen from 3,000 feet with the place name, distance and direction to the nearest airfield. Phoebe and other Ninety-Nines painted more than 16,000 roof tops to guide pilots over the vast American landscape. (When World War II broke out, the markings on the roof tops were painted over to prevent the enemy from navigating the skies over America).

Once women had their pilot's licenses, it was not unusual to find some of them in business for themselves. Some opened flight schools to assist other women who wanted to fly. By far, however, most women in the early age of aviation had involvement through their husbands who owned flying schools, charter services or maintained a Fixed Base Operation (FBO). Because of the joint undertaking, these women drew little attention or fame.

Neta Snook

Neta Snook defied the prevailing social custom and opened an aviation business on her own. Mary Neta Snook Southern was born February 14, 1896 in Mt. Carroll, Illinois and was 95 years young when she took her last flight.

Neta Snook, who had taught Amelia Earhart to fly, had begun flying lessons on July 21, 1917 but had not soloed when civilian flying was banned because of World War I. She remained active in aviation by taking a job with the British Air Ministry, inspecting aircraft engines under production at the Willys Morrow factory in Elmira, New York.[2]

Neta continued her flight instruction after the war. In order to afford flight lessons, Neta worked part-time in a photo shop. The owner would pick up rolls of film from the drugstores in town and Neta would develop and print photos. Neta barnstormed around the country before she soloed for her license.

When she had a total time of 100 minutes of instruction in a plane she helped to build, Neta's instructor and the president of the school decided to take the plane up. The plane crashed, killing the president, seriously injuring her instructor, and destroying the plane. See Fig. 2-2.

Between 1920-1922, Neta became the first woman to operate a commercial aviation field. Located at Kinner Field, in Los Angeles, California, it included passenger carrying, aerial advertising and instruction.

Sunday was passenger carrying day at Neta's airport and one Sunday in December, Amelia Earhart and her father came to the flying field to talk to Neta about giving Amelia flight instruction.

On January 3, 1921, Neta gave Amelia Earhart her first lesson in Neta's World War I Canuck (a Canadian-built training plane). Earhart's first lesson, according to Neta, was for twenty minutes. The first five hours of Amelia's instruction were in the Canuck, and the next 15 hours were in a Kinner *Airster*. According to Neta's good friend Carol Osborne, Neta recalled she had charged

Fig. 2-2 Neta Snook

Amelia $1.00 per minute for lessons. However, the figures in her logbook show she charged Amelia's first 110 minutes at $.75 per minute. Although Neta charged Amelia for instruction in the Canuck, she did not charge her for subsequent lessons in the *Airster*.

Snook's March 30, 1921 records show $85 due her. Apparently they were becoming good friends. Amelia had a box camera and because of Neta's photographic experience, Earhart got an idea to get a part-time job in a photo studio also. Amelia's wages were $21.50 a week at the time.

On January 8, 1921 Neta took the Kinner *Airster* on its first test flight. The flight lasted three minutes (that would have taken her up to circle the field and probably four figure eights). The main problem Snook had was wondering how it would handle on landing. The landing went well and the plane's designer, Bert Kinner was thrilled. After he saw the plane would fly he then went up with Neta twice. (Kinner later sold this plane to Earhart.)

On April 1, 1921 Neta went to her first fly-in. Two actresses, Mary Miles Minter and Ruth Rowland were there. Ruth Rowland wore a black leather jacket, black boots and a black helmet. Neta so admired Ruth's attire that the next day, she used black shoe polish to paint her coat. She also dyed her puttees (calf-length-leather-protectors) black. A month later she painted her Canuck black. The Canuck was still painted black when she sold it in 1922.

On July 3, 1921 Neta took Amelia up for her first 25 minutes of instruction in the *Airster* and two weeks later they had their first crash on Goodyear field.

Kinner had to repair the Airster before Earhart could fly the craft again, and Neta retired from aviation before repairs had been completed.

Neta wanted a child above everything. When she was expecting, she made a vow that if she could have a healthy baby, she would give up flying forever. She had a handsome and healthy son named William Curtiss Southern.

In August 1922 Neta stepped out of her Canuck and never flew again until 1977 (on a commercial flight). She sold the Canuck for a house and lot in Manhattan Beach, and a $500 Liberty bond.

Neta Snook was one of the few people present to see the successful test flight of the Douglas *Cloudster*, the first of the Douglas Commercial series and the first plane able to carry its own weight in payload. Neta went to work for the Douglas Aircraft Company, covering the wings of the first biplane built by Douglas. She sewed the linen covering and helped set up the wing department.

Ruth Rowland Nichols

Ruth Nichols was one of a group of early flyers who dedicated their lives to make America "Air-Minded." She helped to prove that aviation was not just a man's world, by demonstrating that women could do anything in the air that a man could do. Out of the many early records she set came new ideas and equipment, all of which helped to develop the far-reaching industry we know today. Admiral Richard Byrd once said, "There were a handful of women who shared in the hardships and perils of aviation pioneering. Two names that stand out as I look back upon the late 'twenties and early 'thirties, Amelia Earhart and Ruth Nichols." See Fig. 2-3.

Just out of high school, Ruth Nichols took her first flight in 1919. This flight sparked a lifetime passion for flying. After graduation from Wellesley College, Ruth continued flight instruction and received her hydroplane pilot's license on June 4, 1927, the first such license issued to a woman by the Federation Aeronautique Internationale.

By 1926, Ruth had soloed an OX5 "Jenny" and received her pilot's rating in land planes. She was the first woman in New York State to earn this distinction. Eventually she flew every type of licensed aircraft including: dirigible, glider, autogiro, land planes, seaplanes, flying boats, amphibians, monoplanes, biplanes, tri-planes, single, twin, four engine planes and jets.

By 1928, Ruth was co-pilot on the first non-stop flight between New York and Miami in a Fairchild monoplane on floats. The trip took exactly 12 hours and earned her a sales promotion job with Fairchild Airplane and Engine Company.

The same year Ruth became interested in organizing Aviation Country Clubs and during the 12,000 mile Sportsman Air Tour she publicized these clubs. In doing so she became the first woman to complete the tour. Those were the days of uncharted air lanes, few airports, no beacons, radio beams, control

Fig. 2-3 Ruth Rowland Nichols

towers or other modern flying aids. Estimated time of arrival was usually sent by telegram in code. "Arriving G.W.W.P." - (God willing and weather permitting!) In spite of the primitive air conditions and the infant technology of aircraft design, an amazing fact of her traveling to 90 cities in 46 states was that Ruth did not have to make one forced landing throughout the tour.

Ruth was invited to compete in the first Women's Air Derby from Santa Monica to Cleveland, in 1929. At the time there were only nine women licensed as transport pilots and Ruth was one of them. It was August 26, 1929 and Ruth was in third place at the Columbus, Ohio. After having some maintenance done to the plane, she took it on a short test hop. The plane crashed returning from the short test hop.

First contacted by Amelia Earhart in 1927, Ruth endorsed the idea of an organization composed of women who fly. Through occasional meetings and continued correspondence Ruth and Amelia worked out a framework for the women pilots organization. Ruth became a Charter Member of the Ninety-Nines.

After the disappearance of Amelia Earhart, the Ninety-Nines decided to establish a memorial to their first president. In 1939, they established the Amelia Earhart Scholarship fund with Ruth Nichols as its first chairperson.

At this point, Colonel Clarence Chamberlin, who had flown the Atlantic Ocean farther than Charles Lindbergh shortly after Lindbergh made his historic flight, became an influence in Ruth Nichols' life. He invited her to fly with him on the first midnight delivery of newspapers from New York to Chicago. Nichols flew the entire route and Chamberlin later described her pilot abilities. "She has an uncanny feeling for drift and held her course much more accurately, for instance than I do. We started out at night for Cleveland. After she had flown for half an hour, I saw she was so good that I went to sleep and didn't wake up until we had arrived. She is not only a fine pilot but has the rare gift of common sense, the kind of horse sense that means everything in an emergency."[3]

In November, 1930 Ruth Nichols flew from Jersey City, New Jersey to Burbank, California, in 16 hours, 59 minutes, transcending the previous transcontinental women's record held by an Australian woman, Jessie Maude Keith-Miller.

Chamberlin encouraged her to try a solo trans-Atlantic flight. The Crosley Corporation donated a Lockheed *Vega* and before her trans-Atlantic attempt she flew the *Vega* into the record books. In March of 1931, she set a new world's altitude record for women of 28,743 feet, breaking the record mark previously held by Elinor Smith. She followed that victory with a record breaking flight from Newark, N.J. to Washington, DC. This time she broke the record held by Frank Hawks.

By June 1931, Ruth was confident that she could fly the Atlantic. A few woman had made the attempt as co-pilots but all ended in disaster. Only one woman had crossed the Atlantic up to that time and she was Amelia Earhart, who traveled as a passenger. To go alone took nerve, but Ruth had plenty of that. She was piloting her Lockheed *Vega* a powerful plane that required a larger that normal runway for landings and takeoffs. She had been told that the runway at Saint. John, New Brunswick, would suit her requirements. However, when she reached her destination, she immediately recognized the landing field was completely inadequate. Her fuel was running low, and she had to attempt a landing. She overshot the runway and tried to go around. Unfortunately she was not high enough to clear the trees and crashed. The crash left Ruth severely injured with five broken vertebrae. For the rest of her life she required a steel back brace.

The steel back brace did not stop this intrepid woman. In October 1931 Ruth set a new speed record for women by making a transcontinental flight from Long Island to Burbank, California, in 16 hours 59 minutes. On her return flight she was the first woman to set a one-stop transcontinental distance record. She flew from Oakland to Louisville, Kentucky in 14 hours. This surpassed Charles Lindbergh's record made six months earlier. According to the National Aeronautics Association, she was the only woman in the world to have held three maximum international records at the same time, speed, altitude and distance. She also became the first pilot in history to hold these records flying the same airplane.

The following day, October 26, Ruth Nichols was about to take off from Louisville when her plane caught fire and burned. Again Nichols escaped death by tumbling from the cockpit steel brace and all, just seconds before the plane's gas tank exploded.

In 1932, the International League of Aviators awarded Ruth Nichols a special medal as the "Outstanding Pilot in the United States for the Year 1931." This was the first such honor ever extended to an American woman.

Her back injury did not take her drive away and Ruth Nichols still wanted to be the first woman to fly across the Atlantic. However, while she waited for repairs and modifications to be completed on her Lockheed *Vega*, Amelia Earhart made her successful trip.

In the spring of 1932, the New York and New England Airways Company formed with Ruth launching service between New York and Bristol Connecticut. This made her one of the first women airline pilots in the United States.

Ruth went on to set other records. On February 17, 1932, she flew to an altitude of 19,928 feet in a diesel-powered airplane. This was an altitude record for men and women that still stands today.

In 1939, with war on the horizon, Ruth realized that there would be a need for "mercy flying" and air ambulances could do a great service in disaster relief and in everyday civilian emergencies. At the annual convention of the National Aeronautic Association, in Washington, Ruth announced the founding of "Relief Wings." Its purpose she said was to promote a humanitarian air service to provide relief, aero-medical information and facilities for civilian disasters of nature or war

Ruth served as the National director of Relief Wings. With three other women, Ruth set out on a five-month national tour to raise funds and launch the program. This flight was responsible for the foundation of Relief Wings chapters in 36 states. When the war broke out, Ruth turned the organization over to the government-sponsored Civil Air patrol. At the time of her death, in 1960, she was a Lt. Colonel in the Civil Air Patrol.

After World War II, Ruth received an appointment as Special Volunteer Correspondent by UNICEF. She immediately began a round-the-world flight to report the progress in feeding, care and rehabilitation of the world's needy children. Ruth was a "courtesy pilot" that flew a DC-4 Skymaster to Japan, Thailand, India, Pakistan, Iran, Egypt, Israel, Turkey, Greece and Italy. Ruth completed her UNICEF surveys, and departed for home. Unfortunately the plane had to ditch at sea. After a night of fighting mountainous waves most of the passengers and crew were rescued, but not the UNICEF reports. Within 24 hours, Ruth was back in Hartford, Connecticut realizing another dream and setting another record. She was the first woman to complete a globe-circling flight.

In 1958, Ruth was at the co-pilot's controls in an Air Force Supersonic TF-102A Delta Dagger. Ruth took over the controls at 51,000 feet with the plane traveling at over 1,000 miles per hour. She had flown faster than any woman in the world.

Ruth Nichols once reflected: "As I look back on a long series of adventures in the sky, I am grateful to have been able to play a part in the highest drama of our times, the unbelievably swift development of aviation, the new science that has changed the habits of the world and is now setting its sights on outer space."[4]

Amelia Earhart

It started with Harriet Quimby and all the early women aviators were aware that their very presence in the cockpit was crossing the gender line. The women who gained media attention were good pilots. These same women who defied the norms were often criticized for how they dressed as pilots. Some attempted to conform to the dress "codes" expected of women. Quickly they found, as others before them, that wearing dresses, skirts and hats were dangerous, and began dressing for comfort and safety, in men's clothing. Earhart in particular, because of her fame was often the subject of harsh and hurtful comments from her detractors. Some of the comments even arose from her close friends. George Putnam, who later became Earhart's husband even admonished her. "Your hats! They are a public menace, and you should do something about them, when you must wear them at all."[5]

In a newspaper account of a designer line of clothing Amelia was endorsing, she said, "You know, women pilots had a lot of trouble getting practical clothing to wear flying. We had to buy men's clothes or special models made up by stylists and they are seldom really useful or appropriate, besides being expensive. So we banded together and a consultant was appointed by our organization to confer with manufacturers and get something useful in flying togs made.

"I was offered a chance to do some designing and I took it. Years ago in my career I used to make my own clothes from my own designs."[6]

As in many areas of American life, a double standard existed. Men in aviation and other professions were judged on their performance. A man, no matter what he did for a living, was judged primarily on his achievements. A woman, on the other hand, had to be glamorous first, in the pursuit of hers. When Earhart returned to the United States after her trans-atlantic flight in the

Fig. 2-4 Amelia Earhart

trimotor Fokker, *Friendship,* there were thousands of telegrams waiting for her. Most were congratulatory but some admonished her for not combing her hair or for wearing man-tailored clothing. See Fig. 2-4.

Earhart, like many of the women flyers of the Golden Age of aviation, defied other norms as well. Although Earhart started out rejecting financial rewards for her new found fame, the high cost of maintenance soon made her realize commercial endorsements were necessary. One of the pilots on the *Friendship* had carried a pack of cigarettes on the flight and the cigarette company capital- ized on that fact. Although she did not smoke, the cigarette company hired her to appear in their advertisements. Earhart was reluctant since she did not smoke, but if she did not appear she would be denying the two pilots of the *Friendship* their share of the advertisement. Portraying women with cigarettes in 1929 was a bold step forward but the step was not with the intention of liberating them. Earhart was slender and cigarettes were being advertised at the time as a diet aid. "For a slender figure - Reach for a Lucky instead of a sweet," the ads said. Across the front of the cigarette ad Earhart scrawled a sardonic comment, "Is this the face of a lady? What Price Glory!" All but her fans would answer NO because for so many people, it was not "ladylike" for a woman to smoke cigarettes. One letter from a critic concluded, "Since you smoke, I suppose you drink too." Another letter declared, "Cigarette smoking is to be expected from any woman who cuts her hair like a man and who wears trousers in public."[8] When McCalls magazine saw the ad, they canceled Earhart's aviation column (*Cosmopolitan* soon picked up the column). Earhart donated the $1,000 proceeds from the ad to Admiral Byrd's Antarctic expedition and never again endorsed a cigarette ad.

There was a major advantage of dressing in men's clothes as Earhart and other women did. The woman was immediately perceived as a pilot at airports, and not someone's secretary. That in turn gave her easy access to the facilities. On the other side of the coin the implications were more serious. Earhart's more vicious detractors suggested she was a Charles Lindbergh in drag. Worse, a cartoon "Toonerville Folks" by Fountain Fox put it "...and the more pitchers (sic) I see the more I am convinced the feat wuz (sic) performed by Lindy, dressed up to look like a gal."[9] For some insecure and vicious individuals, perceiving Amelia Earhart in this light was easier than having to consider the implications of a woman soloing across the Atlantic or in her the successful performance of other "men's work."

Fay Gillis Wells

Fay Gillis Wells, a junior at Michigan State University entered aviation almost through an ultimatum. She found college boring and dropped out. Her father, however, would not listen to the rebellion of his young daughter. "Either you go back to college, or get a job," he said. She got a job, but after a year was still restless. Fay did not want to return to college, and began to ponder her future. One day she read about the 1929 Women's Air Derby. The article convinced her that women had a future in aviation. She quickly headed for the Curtiss Flying School at Valley Stream, Long Island, New York, to sign up for lessons. Fay took her first lesson on August 6, 1929 and soloed twenty-five days later, after having completed twelve hours of flight instruction. See Fig. 2-5.

The day after she soloed she accepted an offer from her flight instructor to accompany him to test fly an experimental Fledgling Airplane. The machine normally had a 165 horsepower engine, but this machine had a 225 horsepower engine. At 5,000 feet, the pilot began a series of aerobatics. On one maneuver, while attempting to recover from a 3,000 foot dive, the tail collapsed. Then the vibration worked the engine off its mounts and the wings collapsed. The plane, or what was left of it, fell towards the earth. The fledgling student pilot, Gillis, fell out of the disintegrating plane. She was disoriented but managed to open her parachute at 400 feet, and landed safely in a tree. With that exciting adventure behind her, she qualified for the Caterpillar Club.

Fig. 2-5 Fay Gillis Wells

The Caterpillar Club

In the 1920s, the parachute was a device most pilots considered a novelty at an air show. It was something *real* pilots did not wear, only sissies wore them. Hundreds of pilots had also paid the ultimate price for this attitude. However, this attitude began to change when, in October 1922, a parachute saved the life of Lt. Harold Harris. Harris jumped from his disabled plane moments before it crashed, and the parachute he deployed saved his life. His fellow officers quickly presented him with a silk "Caterpillar Certificate" symbolic of the silk used in the early parachutes. Anyone who made an emergency jump from an airplane was eligible for membership, and the Caterpillar Club soon had a growing membership.

On June 25, 1925, Irene McFarland, an aerial exhibitionist, became the first woman to save her life using a parachute. The men in the Caterpillar Club willingly welcomed her into their exclusive club. Gender seemed to be a moot point when it came to a near-death experience.

Undaunted by her near-death experience, Fay went up again, minutes after firemen pulled her from the tree. This time the chief instructor flew her through the same maneuvers to test her courage. A month later, Fay passed her flight test and received her private license on October 5, 1929.

As a member of the Caterpillar Club Fay Gillis earned instant fame in the press. As coincidence would have it, the already famous Jimmy Doolittle had bailed out of a Curtiss Hawk under similar circumstances on the same day. The press started making comparisons, and that earned Fay even more coverage. As a result, the Curtiss company hired her as their first airplane saleswoman. In this capacity she divided her time between sales and demonstration flights at air shows.

In 1930, Fay earned her commercial license and sailed off to Russia to join her father who was building electrolytic zinc plants. Fay's bail-out at Curtiss Field had earned her a reputation that followed her around the world. Fay was fluent in Russian, and the Russians invited her to lecture on the art of parachuting at the Moscow Aviation Institute. Letting Fay Gillis fly in Russia was another story. It took her nearly two years to get the permission the Russians had earlier promised when she applied for her visa.

Fay spent four years in Russia lecturing and working as a correspondent for the New York *Herald Tribune,* and the *Associated Press.* She even became America's unofficial greeter of her fellow countrymen who were attempting to capture new around-the-world records. She spent weeks preparing for Wiley Post's flight through Russia.

Gillis continued to fly around the country, and her prestige and perseverance began to build among the Russians. They eventually gave her permission to fly a glider and when her craft arrived, she soloed it without a single lesson.

Fay returned to the United States in 1934, to become the fashion editor of *Airwoman,* the publication of the Ninety-Nines.

Wiley Post called Fay in 1935 and asked for her help in obtaining a Russian visa for him and his wife. Post invited Fay to join them on a pleasure trip across the Bearing Sea. She jumped at the chance. While Post was getting his Lockheed Orion airworthy for the trip, Gillis eloped with Linton Wells, a foreign correspondent she had met in Russia. The two agreed that Fay would still accompany Post but a month later her new husband received an assignment to cover the Italo-Ethiopian War. Linton offered Fay the option of continuing her plans to accompany Post and his wife or join him covering the war. Post assured Fay that he could find a substitute easily, and Fay and her husband departed for Africa. Later Fay learned that the famous American humorist Will Rogers had joined Post and both men died when the plane crashed on take off from Alaska. "It just wasn't my time," Fay reflected.

Fay Gillis Wells covered the war as a correspondent for the New York *Herald* and was eventually arrested and labeled by a British newspaper as a spy. Fay returned to the United States in 1936 and continued to work for the *Herald Tribune.* In 1942, she and her husband returned to Africa to serve as Chief and deputy chief of the U.S. Commercial Company, the trading arm of the War Production Board. During the war she arranged for the purchase of strategic war materials for the United States.

Fay and her husband returned to the United States in 1946 to raise their son. Fay came out of retirement in 1963 to become the White House correspondent for Storer Broadcasting Company. She covered the travels of Presidents Lyndon Johnson, Richard Nixon and Gerald Ford, making two trips to Viet Nam with Johnson, and Nixon and accompanying Nixon to China in 1972.

Fay was instrumental in the lobby effort to create the Amelia Earhart stamp, the first stamp honoring women in aviation. The stamp was approved in 1963. As chairman of the Ninety-Nines International Bicentennial Celebration in 1973, Fay was also instrumental in having hundreds of Ninety-Nines fly seedlings from all the state capitals to publicize the creation of the International Forest of Friendship in Atchison, Kansas, the birth place of Amelia Earhart. The forest, a gift from the Ninety-Nines and the city of Atchison, was dedicated on July 24, 1976, on what would have been Amelia's 79th birthday.[10]

Evelyn "Bobbi" Trout

Evelyn "Bobbi" Trout was born in Greenup, Illinois on Jan. 7, 1906. During her early years, she and her younger brother Denman "Denny" Trout moved around quite a bit before settling in Southern California during 1920.

Evelyn earned her nickname "Bobbi" after having her hair bobbed short.

From the day Bobbi saw her first airplane, she just knew she would learn to fly: it was in her blood. Bobbi was 16 when she had her first airplane ride, in an OX5 powered Jenny. By the age of 22 she earned enough money to make her dream come true. She went over to Burdett Fuller's Flying School in Los Angeles and signed up for a flying course. Her first lesson was a disappointment to her. It was on paper, was on the principles of flight. Her next lesson, however, was in the sky, on Jan. 1, 1928, and that started a long and exciting career in aviation.

Bobbi learned three rules. She can still recite them today: "As soon as you cut power, push the engine down so you don't spin; keep looking for a place to land; never come in low because of high tension lines." See Fig. 2-6.

One day her part-time instructor showed her how to set down in a hay field, violated rule number one and spun-in, almost killing both of them. After the required hours in the air, Bobbi got her transport license. She was the fifth woman in the USA to obtain this rating. She became a demonstration and test pilot for R.O. Bone of the Golden Eagle Aircraft Company during the fall of 1928. She flew his 60 horsepower Golden Eagle on demonstration flights.

On Jan. 2, 1929, Bobbi established a World's Record for Solo Endurance Flight for Women of 12 hours and 11 minutes, 37 seconds She also set the record of being the first woman to fly all night. On June 1, 1929 Bobbi went on to establish a new women's altitude record of 15,200 feet in a 90 horsepower Golden Eagle Chief from Grand Central Air Terminal, Glendale, CA.

Bobbi is a Charter Member of the Ninety-Nines, and she flew in the first 1929 Women's Air Derby (dubbed the "Powder Puff Derby" by Will Rogers) from Santa Monica to Cleveland, Aug 18-26, 1929. Bobbi flew her Golden Eagle

Fig. 2-6 Bobbi Trout recently in her red Porsche

Chief, 100 horsepower Kinner engine, but ran low on fuel and made a forced landing on the approach to Yuma, Arizona, on the second day of the race. Repairs required about three days and Bobbi trailed the race leaders. She spent the night of the 23rd at Pecos, Texas while the others were in Wichita, Kansas. Bobbi did manage to arrive in Cleveland in time for the banquet on the evening of the 26th.

In 1929 Bobbi made several in flight refueling attempts. After five attempts and five re-riggings, she succeeded in establishing a record, Nov. 27 through 29th. Bobbi and Elinor Smith stayed up 42 hours, 5 minutes and refueled 3 1/2 times. This was the first refueling endurance record of its kind for women. The support plane's engine (a Pigeon) fell apart during the 4th contact and with no other refueling ship back-up, Bobbi and Elinor had to land. After this flight Dr. Burton R. Charles made her a leather, electrically-heated flying suit. He invented the suit to protect aviators from the cold when flying at high altitudes. Anne and Charles Lindbergh had their electrically-heated flying suits fabricated after seeing Bobbi's suit. Bobbi later donated the suit to the Los Angeles County Museum of History and Art.

May 30-31, 1930 was the Official Opening of the United Airport (United Air Terminal). Pratt & Whitney sponsored the air race and there was a women's closed course air race around pylons. Bobbi won the race that day flying a Kinner Fleet. Mrs. Hamilton of Hamilton Propeller Company presented a silver cup to

Bobbi for winning the air race in competition with Pancho Barnes and Margaret Perry.

Bobbi created another refueling record with Edna May Cooper, from January 4 to January 9, 1931. They stayed in the air for 123 hours, 50 minutes, until their Curtis *Robin* engine developed a cracked piston. In spite of the difficulties they made eleven refueling contacts during the 5-plus day flight. They had planned on at least 30 days aloft. Bobbi was planning a Trans-Pacific flight from Hawaii to the Mainland (Los Angeles) during the summer of 1932 (August 1-7, 1932) in a modified Lockheed *Altair*, but the flight was canceled because of the Depression and the lack of financial backing. Had Bobbi been fortunate enough to have George Palmer Putnam as her promoter, she feels she would have made this flight. Amelia Earhart accomplished this historic flight on January 11-12, 1935.

Bobbi taught a friend, Mary Wiggins how to taxi, takeoff, and land, in a Cycloplane. Bobbi regulated the amount of gas and Mary would venture a little farther from the ground each time. Soon she was flying.

A May 10, 1937 Los Angeles *Examiner* article announced that Bobbi and four other women flyers were made Aero Police Women. They were Edie Curtiss, Smitty Smith, Ruth Vargas, Peggy McLean and Patty Willis. Mary Charles promoted the idea with the L.A. Chief of Police Davis. Bobbi thinks the intent was to help in an emergency, but it was more honorary than anything, as they never did do a job for the police. They had the privilege of using the Police Academy firing range where Bobbi spent considerable time practicing with a used .38 caliber revolver she purchased from an officer. Not long after she had practiced shooting, she won second place in a match for women. The food at the restaurant was so delicious Bobbi often appeared at the academy for practice around noon (and for lunch). The Aero Police Women had regular police badges with the words AERO POLICE on them, and Bobbi was number four.

"I wanted to become a Police woman during the depression," Bobbi said, "but I could not get in as I was underweight. Those were the days when flying did not pay much and many of us were flying free just to keep up our licenses. I remember Pancho Barnes had her lovely home filled with unemployed flyers."

In the 1930s, Bobbi became active in the Women's Air Reserve (W.A.R.) (sic.) which was headed by Pancho Barnes. "One day Pancho asked me what I wanted to be, I said captain, so she made me a captain." Captain Bobbi Trout was a hard working member of the W.A.R. from 1931-1941. The organization was preparing for national emergencies and disasters. Trout was a member of the Civil Air Patrol (CAP) in the first part of World War II. Her squadron commander was actor Bob Cummings.

During World War II, Bobbi and her business partner, Pop Lingerfelt, invented and patented a group of rivet sorting machines. Based on these machines they started the "Aero Reclaiming Company" which employed as many as 50 people at one time. This was the most complete rivet sorting business in the country. Douglas Aircraft in Santa Monica was the first customer.

Trout sold her interest in the rivet sorting business when she found out that the aircraft companies needed a better way of de-burring aircraft machined parts. Bobbi rented a cement mixer and put it in the back yard. She obtained some parts that required deburring to try a new method of deburring by tumbling. Based on its success she and her new partner rented a building on Flower Street in Glendale and started "De-Burring Service."

In 1976, the OX5 organization named Bobbi, "OX5 Woman of the Year," an honor for any aviator. To commemorate the 50th Anniversary of the 1st Women's Air Derby of 1929, she flagged off the Air Race Classic in 1979, and the Angel Derby the same year. Both began from Clover Field, Santa Monica Airport.

Bobbi Trout ended her aviation flying activities on February 11, 1984, flying Jim Nissen's OX5 Jenny as she celebrated her 55th Anniversary of the "First Woman to Fly All Night" on NBC Television in San Francisco. Jim took Bobbi up and let her take the controls of a Jenny for the first time since 1929.

Florence Lowe "Pancho" Barnes

One of the most colorful characters in aviation history was "Pancho" Barnes. Florence Lowe Barnes was born on July 29, 1901 and she lived at full throttle until she died in 1976. Florence was born into a prominent Pasadena family and fortune. Her grandfather, T.S.C. Lowe, was the founder of Cal Tech. He also built Mt. Lowe Railway and the observatory on Mt. Lowe. In 1856, he constructed balloons to study atmospheric phenomena. In 1858, he built and operated the largest aerostat up to that time. (Lowe made his fortune in gas as a serendipity of ballooning.) During the Civil War Lowe was chief of the Aeronautic Corps. He organized the first aerial military war unit in the world, which is now the United States Air Force, and was its first commander-in-chief. See Fig. 2-7.

Pancho Barnes spent her childhood in Pasadena. Her first school was an all-male elementary school across from Cal Tech. From there she went to a school for girls and then on to the Ramona Convent. She rubbed elbows with Hollywood movie stars, moguls and the intellectuals of the day. But she shunned the glitter and went on to earn her fame in a field that normally did not welcome women. She became famous for both living as high as she flew and dissipating the family fortune in the process.

While Pancho was still a young girl, she married an episcopal clergyman in Pasadena. "This being married to a minister was a little too slow for me," she said years later.[11]

"I just didn't stay home very long." Pancho traveled constantly with horse shows, and in 1926, went to South America for a year. After she tired of South America she traveled through the West Indies.

She traveled alone and in 1927, signed on a banana boat, as a crew member. That boat, she found out after signing on, was also running guns to Mexican revolutionaries. When she got to Mexico she found herself in the middle of a

Fig. 2-7 Pancho Barnes

revolution. It was there she earned her nickname "Pancho." When the shooting got too close for her (a man, standing beside her in the street was shot dead) she decided to hobo across Mexico and ended up on the seacoast where she got a job on an American vessel. She cut her hair, wore men's clothes and successfully disguised herself as a seaman.

In a 1969 interview she was asked what her husband thought of her travels. She replied, "I don't know...I didn't worry about what he thought."

In 1928 Pancho started flying. "A fellow named Ben Chapman taught me to fly. He charged $5.00 for a fifteen minute lesson. He didn't teach me much, the only instrument I had was an oil gauge." Over the years, Pancho worked as a test pilot for Bach, for Lockheed and, for Beechcraft as a production pilot. She also went to work for Union Oil Company in 1929, and flew for them three years. There she did air racing, and promotion. "I was a stunt pilot then. In those days nobody knew what they were looking at so we did anything we pleased. Anything you did was just great." At the beginning of the Depression the pilots who worked for the oil companies had good jobs and received large sums of money. By then, Pancho had her own planes, and one of them - the "Mystery Ship" - was the fastest plane in the world.

She signaled her rejection of her marriage to the minister by buzzing the church during his Sunday sermon. The entire service was disrupted by her low pass. The couple divorced in 1942.

For Pancho Barnes, flying was life's ultimate high. Someone once said of Pancho, "Flying made her feel like a sex maniac in a whorehouse with a credit card."[12]

On August 5, 1930, Florence "Pancho" Barnes established a world speed record for women. Leaping into the cockpit of her hot, new $16,500 Model S Mystery Ship, which Travel Air built especially for her, she flashed low across Los Angeles' Metropolitan Airport in four breathtaking sprints. Officials clocked her black-and-red monoplane at 196.19 mph, a record that stood until Ruth Nichols hit 210.636 mph the next year at the Detroit air races. In doing so Pancho broke Amelia Earhart's record of 181.8 mph set a month earlier at Detroit, Michigan.

Pancho even flew movie stunt scenes, flying in such pot-boilers as "The Flying Fool," "Sky Brides" and "The Lost Squadron." She tired of movie-making though, and turned to politics, unsuccessfully running for county supervisor from the Third Los Angeles District.

Pancho admitted to being an opportunist. "I was one of two women who was getting out there and really flying. I was the only movie stunt pilot they had. I worked in Howard Hughes' Hell's Angels and all those other pictures. I was the only woman around who could do that kind of work."[13] She also created a storm of controversy around herself but she successfully organized the Hollywood Stunt Pilot's Union.

By 1933, Pancho Barnes was broke. She had gone through the family fortune, with the help of the Depression, and had to sell her apartment in Los Angeles, and her 30-room mansion in San Marino. She used the money to buy 80 acres of desert farmland in California. By 1947 she had parlayed the 80 acres into 380 acres of land. See Fig. 2-7

In 1934, Pancho organized a memorable cross-country flight to win a presidential pardon for Duncan Renaldo, a handsome movie star who had stumbled into trouble. Renaldo was behind bars at McNeil Island Federal Penitentiary, put there by the Immigration Service for failing to sign his real name, Vasile Dumitree Cughienas, to his passport.

Pancho went into action on Renaldo's behalf by organizing a squadron of women pilots that included Viola Neal, Bobbi Trout, Nancy Chaffee, Mary Charles and Patty Willis, all dedicated fans of the handsome movie star. They swarmed east to Washington and there put on a memorable air show that so delighted President Franklin Roosevelt that he turned Renaldo free.

Back in Los Angeles, Renaldo demanded to know who had arranged his premature release. "Shucks, Renaldo, I did," said Pancho. He kissed her.[14]

In 1934, Pancho Barnes organized the Women's Air Reserve. Its primary purpose was to provide aid in national and community emergencies. Its members consisted of pilots, nurses and doctors.

Pancho has been described by those who loved her as, "A mile and a week less than beautiful." Test pilot and close friend Chuck Yeager, who gave her away at her fourth wedding, lovingly described her in his autobiography, "She had to be tough, because the only argument about her was whether she was the ugliest woman in the world or just one of the ugliest." Her last husband, Eugene "Mac" McKendry dismissed those comments. "I could see an inner self," he said.[15]

Despite her heavy, moon-shaped face and lifelong disdain for dresses and other things "conventionally female," Barnes bragged, Yeager writes in his book, of her long list of lovers. She claimed to have had as lovers silent screen star Ramon Navarro, and stunt pilot Frank Clark, both of whom, like her fourth husband McKendry, were very good looking. Personality, intelligence, boundless energy and generosity for her friends were her charisma. "She was a woman ahead of her time," said McKendry. "She was always one of the boys because she thought of herself as equal to men and should be able to do everything they did."

In the late 1930s, the United States was preparing for war and Pancho's ranch in Antelope Valley, California looked good to them. The government asked her to start a civilian pilot training school on the ranch to which she willingly agreed.

Not only did she turn out fledgling airmen under the Civilian Pilot Training Program on the ranch, when the Army Air Force got caught short of manpower at the onset of World War II, but she gave the GIs at her ranch the kind of morale boosting they really understood. "I had a three-ring circus going out there," she said. "The riding club started during the war. The boys on the base had nowhere to go, so we rode horses during the day, and had big meals at night. We got liquor where we could and set it out on the tables and punched cards for drinks."

"I just hired the dolls to serve booze, and I laid down strict ground rules for their behavior."

The "Happy Bottom Riding Club," as the ranch became known, was an unbelievable institution that encompassed a private airport, bar and dance hall peopled with handpicked, buxom, sun-browned young women.

During the war years, Pancho kept her mouth shut and prospered by creating a sort of perpetual motion machine, "When I started the guest ranch, I had the milk contract for Edwards Air Force Base, (then Murock), for the Navy base at China Lake, and for the Marines at Mojave. I sold all the milk they used and I collected all the garbage." She charged the air base to collect its garbage, then fed it to her 3,000 hogs and then sold the air base the ham to accompany their powdered eggs.

Pancho Barnes had three more marriages, one to a flight student which lasted just weeks, one to a magician which lasted less than a year and the last to Eugene (Mac) McKendry. That marriage lasted almost 15 years.

The year was 1946 and "Mac" had just come home from a tour of duty with the Army Air Force Transport Command as a "Hump" pilot. He had also been greeted with divorce papers by his first wife. Mac was 26, and Pancho was 44. McKendry remembers the first time they met. "Pancho was wearing boots and a suede jacket with fringes. I thought; There's someone who needs my help. I think we both needed each other. I really do, and we adopted each other from that day on." [16]

When Pancho married McKendry, on June 30, 1952, the operations at Edwards Air Force base came to a standstill. She invited the whole base to her wedding. The champagne corks had barely begun to pop when a monster B-47 Stratojet thundered over the ranch, slow-rolled off into the distance and landed at Edwards. The pilot was General Boyd, who had just completed a record run from Dayton, Ohio. He drove to the ranch just in time to join Chuck Yeager in giving the bride away.

Fifteen hundred friends of Pancho watched a full-blooded Indian, Chief Lucky of the Blackfoot tribe, perform the marriage ceremony.

Reflecting on the ranch, Pancho said, "After a while, this place really started jumping, but there was such a predominance of men that I made it a point of hiring hostesses. They would come and dance and after a while the place just grew and grew. We had a dance hall motel and live bands two or three times a week. It was a real jumping joint."

The club had as many as 400 guests for the Wednesday night dance during its prime in the late 1940s and early 1950s. There were underwater ballets in the club's pool featuring nude swimmers. [17]

The club had become a hangout for pilots like General Jimmy Doolittle, and Chuck Yeager. They were Barnes' first official members when, in an attempt to control membership, she made it into a "private club." Doolittle is credited with coming up with the name, after a long ride on a spirited horse. He said the ride gave him, "a happy bottom."

By the 1950s the club also began to bother the Air Force brass; it sat directly in line with a projected twenty-seven-mile runway scientists thought would be needed to recover outer space vehicles. The ranch had to go.

The quickest way to eliminate Pancho's ranch, they decided, would be to dub The Happy Bottom Riding Club a house of prostitution and Pancho its madam. This would give them a good excuse to declare her out of bounds.

In an unparalleled action in Los Angeles Federal Court, Air Force legal experts sought to shoot down Pancho with one salvo by proving that she was dangerously undermining the morals of the military personnel at the base.

Pancho already had drawn first blood in the bitter legal dual; her thrust was a suit for damages over her alleged status as a madam.

Disdaining legal advice, Pancho appeared in court *in propria persona,* acting as her own attorney. The lady of the dry-lake resort played the part of the downtrodden individualist, up against the "system." She came to court dramatically dressed in cowboy boots, riding britches and colorful western shirts from Senator Barry Goldwater's exclusive cattle-country dude store in Phoenix.

Her appearance was designed to provoke an entirely different image than the gossip had created. She had piercing, wide-set eyes blazing from a suntanned face, softened by wispy, dark-brown hair. The years of outdoor living had left her body physically trim, and she definitely did not appear the type-cast night-crawling bordello operator.

Pancho said what people did behind closed doors was their own business. She made the point not indelicately with a "Notice of Non-Responsibility," in a printed sign displayed over her bar: "We're not responsible for the bustling and hustling that may go on here. Lots of people bustle, and some hustle, but that's their business, and a very old one."

Her critics maintained it wasn't the principle of the thing, it was the money involved that touched off her celebrated court battle with the Air Force.

"Nonsense," she said. "They just picked on the wrong girl to push around."

In the federal building press room, one veteran reporter whistled when he read the complaint she filed.

"You don't actually expect to collect $1,483,000 do you, Pancho?" he asked. This was the price tag for her damages and maligned hog ranch, which the government had set out to buy for much less.

"Hell, no!" she snapped. "But I'm going to worry 'em to death!"

Her second blast came on December 3, 1952, when she filed a new complaint seeking another $1,253,546.29 in damages for a variety of acts from conspiracy, harassment, fraud and deceit.

The way she told it, General Holtoner had warned her that somebody might accidentally drop a napalm bomb on her ranch on take-off from the new runway if she didn't get out. She slapped Holtoner with a $300,000 suit on the contention the warning constituted a threat to bomb her off the desert.

Nobody actually bombed Pancho's ranch from the air, but on November 14, 1953, a mystery explosion knocked out one end of her dance hall and started a fire that leveled the structure and gutted her eight-room ranch home. Only a few charred mementos escaped damage.

In 1954, a federal judge ruled against Pancho in her claim she was the victim of a conspiracy by the Air Force to ruin her business but ruled also that the prosecution had not proven she was a madam either.

When the smoke of the appeals cleared in the early summer of 1956, Pancho picked up her check for $414,000 and said good-bye to The Happy Bottom Riding Club.

The question, "Was she or wasn't she?" was never cleared up legally. But does anyone really care?

A writer for the *TV Guide* said just before the 1988 screening of the made-for-TV movie, *Pancho Barnes*, "The screenwriters had a time turning an "R" rated life into a prime time movie."[18]

Pancho Barnes died in 1975 and every September thousands of people attend the Pancho Barnes Memorial Barbecue at the site of her old ranch. A band plays on what was the foundation of the clubhouse and the proceeds go to fund a Test Flight Center Museum at Edwards Air Force Base where her memorabilia is displayed.

Janet Harmon Waterford Bragg

Black women aspiring to pursue careers in aviation during the first half of the nineteen hundreds encountered seemingly insurmountable obstacles. It was an era when many considered blacks mentally unqualified to become pilots. That perception did not stop Janet Harmon Waterford Bragg from following her dream to fly.

Janet was born in Griffin, Georgia in 1912, as the youngest of seven children, and grew up a tomboy in the shadow of her four older brothers. She graduated from high school in 1927 and attended Spellman College in Atlanta where she earned a degree in nursing. She moved to Chicago to start her career but flying remained a fascination.

One day in 1933 she saw a billboard displaying a bird building a nest around her chicks. The caption read: "Birds learn to fly, why can't you?" That incident clinched it for Janet and she promptly enrolled in the Curtiss Wright Flying School in Chicago. She was the only girl in a class of 26 boys but studied right along with the fellows, learning everything from aerodynamics and meteorology to powerplants and navigation. "The ground school was interesting but I couldn't wait to get up in the air. I finally asked the instructor, "When do we get to do some flying? He laughed and said whenever you're ready.

"I'll never forget my first flight instructor, his nickname was 'Dynamite' and he was a real roughneck. He took me through all sorts of maneuvers on that first flight. I think he thought if he could frighten me or make me sick I'd get scared and never come back, then he'd never have to deal with me again. But I sure fooled him, I loved the ride and couldn't wait to go up again."

Janet continued taking private flying lessons but the $15.00 an hour rate proved to be too costly. The obvious alternative for Janet and her flying friends was to purchase an airplane of their own. Janet used her nurse's savings of $600 to buy an OX-5 International. She later commented, "It was all the money I had at the time but I'll never regret spending it on that airplane." Janet and her friends created their own flying club called the Challenger Air Pilots Association and the members immediately elected Janet as their first president.

Buying the airplane turned out to be easy compared to the problems they faced finding a place where they could fly the machine. Few air fields would accept blacks so Janet and her group cleared their own runway when the village of Robbins, located southwest of Chicago, agreed to give them a piece of land. "We were restricted from flying over the houses and we abided by that but folks still complained. Nearly every Sunday the police would come out and arrest one of us for making too much noise."

It was a rough time for the Challenger club and Janet did what she could to keep her group flying. Besides providing encouragement and support she devised various fund raising projects to defray flight expenses and to keep flying affordable.In 1939 the members of Janet's group formed the Coffey School of Aviation and they were allowed to participate in the Civilian Pilot Training Program. The program was designed by the government to feed students into the Army Air Corps training program in Tuskegee, Alabama.

When the United States entered World War II, American men marched off to war and those left at home answered their country's call to serve where they could. Janet learned about the WASP program and submitted an application with her flight qualifications. A telegram soon arrived inviting Janet to attend an interview at the Palmer House in Chicago and she was filled with anticipation.

Janet's arrival stunned those in charge as they had no idea that black women were flying. The startled interviewer asked, "My dear do you fly?" and remarked, "Why I've never interviewed a colored girl before." She questioned Janet about her training and admitted she would have to get authorization before accepting her into the program. Jackie Cochran sent a reply thanking Janet for her interest, but with segregation the norm in the South, and the WASPs housed at the training barracks in Sweetwater, Texas, there would be no place for her to stay. Janet applied further as a flight nurse, only to be told that the quota for "colored nurses" had already been filled. See Fig. 2-8.

Janet went on to Tuskegee, Alabama and received advanced flight instruction from Charles "Chief" Anderson, a well known instructor in the black flying community. In 1946 she purchased another airplane, an L-235 Super Cruiser, which she used in her cross-country flying to qualify for her commercial pilot's license.Though she was scolded for entering the testing center from the"Whites Only" door, she passed her commercial written exam with a near perfect score. Her flight test the following week would be the deciding factor. "The air was as smooth as silk when the examiner came out for the checkride. The flight lasted a little more than an hour and I preformed all the maneuvers perfectly, including the forced landing test."

When the flight test was over Chief Anderson anxiously waited to hear the results and asked the examiner how Janet had done. The examiner replied, "I'd put her up against any of your flight instructors, but I've never given a colored girl a commercial pilot's license and I don't intend to now." Janet noticed the tear that rolled down Chief Anderson's face but accepted the examiner's decision graciously. She had come a long way and had overcome many obstacles to experience the discouraging setback but she stood tall in the face of adversity. "I was taught that you couldn't hate people for the things they did. People's

Fig. 2-8 Janet Harmon Waterford Bragg

actions are based on things they believe in and I accepted that.I told the examiner, "that's alright, I understand" and thanked him for his time." Janet continued flying and was successful in completing her commercial pilot's license on the second try later that same year.

Janet bears no malice for the discrimination she experienced in her struggles to become a pilot and instead believes that the battles were well worth the fight to fly. She identifies with the bumble bee an insect, who on paper is aerodynamically incapable of flying, but in reality flies merrily along. In a way, society placed its aerodynamic limits and restrictions on Janet and her ability, but like the bumble bee, she never let it stop her from completing her flight plan.

The flying exploits of Janet Harmon Waterford Bragg can be found permanently displayed along with her personal flying paraphernalia at the Pima Air and Space Museum in Tucson, Arizona. At 80 years old, she still encourages the youth of her community to follow their dreams and is a charming and entertaining lecturer at national aviation events.

Chapter Three

Women of Courage

WASP accomplishments formed the building blocks of the present. We know the future is predicated on the past - more than ever the doors for women have opened wide. Pick your goal, and do it! An aviation career for a woman today presents such an exciting world - truly around the world, above and beyond. Kay Brick - WASP

When the United States Air Force Academy opened enrollment to women in 1976, the Air Force also announced that women would begin pilot training. Most of the public and the media reacted as though both events were something that had never happened before. They were right about women being admitted to the Academy, but not about women flying military aircraft. The women who flew military aircraft for the WASPs during World War II received little publicity and disappeared into the dark recesses of history for three decades after the war.

A Crucial Role

During the war years, 15 million men went into uniform. The nation had to fill the vacuum created by the men going off to war and called on the women of America. Within a year, there were over half a million women working in the aviation industry, filling the ranks as engineers, specialists, factory workers, etc. Roles were changing and women volunteered to meet every wartime demand. By late 1943, women in the aviation industry skyrocketed to more than 36 percent of the industry's work force. The second World War was also the first opportunity for women to play a crucial role as pilots. World War II would leave a permanent change in the social fabric and women's role in America. Gender lines separating women's work from men's work in industry would begin to blur, and heroism and bravery would no longer be exclusively reserved for men.

To better understand the barriers facing women in the early 1940s, one need only look at the U.S. Census statistics. Only a small portion of America enjoyed flying for sport or as a livelihood. There were 130 million Americans in 1939, but only 30,000 of them possessed a pilot's license, and 13,000 aircraft compared to more than 30 million motor vehicles. While 30,000 licensed pilots may seem small to some, the representation of women in those ranks was minuscule. Four hundred women belonged to the Ninety-Nines, the International Women Pilots Association, and 160 women had commercial licenses.[1] Forty-three women were listed with the CAA as instructors.[2]

At the start of the war, there were only small numbers of women employed in the Boeing, Douglas and Lockheed aviation plants. Fewer still were at the level of maintenance and air traffic control. In 1942, Elsa Gardner was the only female aeronautical engineer in the Navy. Isabel Ebel was another engineering pioneer. She graduated from MIT in 1932, the only woman studying aeronautical engineering among a student body of 30 women and 3,000 men. As a woman she found it impossible to find work in any American aircraft factory, so she went back to school at the Guggenheim School of Aeronautics at NYU. No woman had ever been admitted, and it took the intervention of Amelia Earhart to get Ebel accepted.[3]

Besides the high profile flyers like Earhart, Cochran and a few others, aviation as a livelihood was still a man's game. Those women that did fly had several unique qualities. They usually were well educated although there were exceptions. Some like Jacqueline Cochran had wealthy husbands and others had independent incomes that could support the expenses of lessons and airplane maintenance. Others worked two jobs to fund their love of flying. Those women who entered the male-world of aviation were generally accepted after demonstrating their ability because they brought with them a self-assured quality.

Three outstanding women who worked with the aircraft industry during the war were Cecil "Teddy" Kenyon, Barbara "Kibbee" Jayne, and Elizabeth Hooker. They were hired to be test pilots by "Bud" Gillies, a vice-president of Grumman Aircraft Corporation and husband of Ninety-Nines president, Betty Gillies.[4]

They flight tested every airplane as it came off the Grumman assembly line. These pilots were the first to fly the aircraft, and were exposed to a high degree of risk in order to certify the aircraft for delivery to the military.

Despite the excellent work of Kenyon, Jayne, and Hooker, there was anxiety at Grumman. Margot Roberts described this problem in her 1944 article in *Woman's Home Companion*:

As a class, men fliers are nobly [sic] conservative, masculine and inclined to think of a woman's place as being in the home. They rarely welcomed lady pilots in their midst. This is something that all the gal pioneers of flying have bumped into at times. As they have the normal feminine desire to get along with men, a problem is created at first - though a very short-lived one in most cases - and you can't always blame the men. There are several angles to this man-woman equality proposition. At Grumman, for example, there is the matter of sticky ships. . .

"Sticky ships" were airplanes suspected of having serious problems that could not be found by mechanics on the ground. The women were not permitted to fly these aircraft, nor were they permitted to do experimental testing such as dives. The three were annoyed by the arbitrary and capricious policy. It also complicated their acceptance by their male colleagues. Jacqueline Cochran was also critical of the job the Grumman women were doing. She called it a kind of "aerial dishwashing," yet within the restrictions imposed on them, Kenyon, Jayne, and Hooker had ably demonstrated the competence of female test pilots. As Margot Roberts put it, "It isn't real dishwashing and one reason for the attraction."[5]

British Air Transport Auxiliary (ATA)

In 1939, it was clear that America was going to be involved in a global war, and women could play a major role in the country's defense. What was not too clear was when the United States would enter the armed conflict already raging in Europe. It was also obvious that the air would play a major role in the war. Plans and estimates for pilot requirements needed to be in place before the United States was actually drawn into the conflict. Isolationist factions versus militarism went head-to-head and by 1941, the nation still had fewer than 100 battle-ready aircraft, an Air Corps of 17,000 men and a capacity to train only 700 pilots annually.[6]

When Franklin D. Roosevelt took office in 1933, he provided the American public with a refreshing point of view. His administration was remarkable from the standpoint of women's rights. He elevated to national status a select group of women reformers and activists. Some women found themselves with access to the highest levels of political power. There were two reasons for this. One, their skills, abilities and contacts outside politics made them a natural match for the objectives of the "New Deal." The second reason was FDR's wife and political confidant, Eleanor who turned the position of First Lady into an active and intellectually vigorous power center.[7]

Mrs. Floyd Odlum, (AKA Jacqueline Cochran) had several excellent connections in government. She was aware of the strategic need for pilots when war came. Floyd Odlum was a friend of the Roosevelts (and many other prominent politicians) and a major contributor to Roosevelt's political campaigns. In September 1939, Cochran wrote a letter to Eleanor Roosevelt suggesting the use of women pilots for non-combat jobs in the upcoming war. She had a plan to recruit highly qualified women pilots to assist the Army Air Corps in the non-combat flying jobs. Eleanor Roosevelt was enthusiastic about the idea. Cochran maintained that the real bottleneck in the long run was likely to be trained pilots. She proposed that male pilots could be released for combat duty by assigning women to their former roles. But as much as the President listened to the advice of his wife, he was also getting advice from men like General George Marshall, and General Henry Arnold and they did not, at the time, think such a plan was necessary. It would take time for the idea and the need to come together.

Cochran had served on the Collier Trophy committee in 1939, and after the presentation made by President Roosevelt at the White House she met with General Henry H. "Hap" Arnold and Clayton Knight, the acting head of an American recruiting committee for the British Ferry Command. During lunch, Knight discussed the dire need for pilots in England. Knight asked Cochran to help recruit female pilots for England. When General Arnold suggested that Cochran actually do some of the flying, Cochran immediately saw an opportunity. She then began to put the plans in place to select women for the British Air Transport Auxiliary (ATA).

Other Programs

In the meantime, several civilian and quasi-military organizations formed to tap into the potential women had to aid in the war effort, specifically in aviation. Many women who joined these organizations and became pilots would never have been selected in a formal military program because of their gender, martial status or children. These organizations provided some women with their only chance to become involved in aviation during the war.

An organization formed by women called the Women Flyers of America (WFA) proposed to address the shortage of women pilots in America. They set up courses approved by the CAA at rates about 20 percent less than the prevailing rates. They were not prepared for the response. Within a week of the announcement in April 1940, over 1,000 women applied. The response immediately launched the WFA into a national organization. The applicants for the WFA were young, adventurous women between the ages of 20 and 35 years, and the WFA afforded them a way to learn to fly and help the country in a patriotic way. The women saw their roles as ambulance and liaison flyers when war came.

Another program to encourage Americans to learn to fly was the Civilian Pilot Training Program (CPTP). Unlike the WFA, it had a major restriction. It admitted only college-aged students and had slots for about one women for every ten men.[8]

As the war clouds got closer, the CPTP took on the character of a military organization. In early 1941, the trainees signed a nonbinding agreement that in case of war or national emergency, they would serve in the military. The women even filled out cards specifying the branch of service they preferred, but in June 1941, when the nonbinding agreements became a legal obligation, the women were instantly excluded from the CPTP. The women protested and their strongest supporter, Eleanor Roosevelt publicly demanded an explanation. She received a reply from the CAA that in effect said men had a wider potential for usefulness in the armed forces as pilots. They would concentrate on this resource, rather than split the CPTP resources between men and women.

Some individuals in the CAA had strong feelings that women could best serve the war effort as instructors. In the summer of 1941, Phoebe Omlie, now living in Tennessee with her husband, opened a school to teach young women to become flight instructors. A Tennessee aviation gas tax funded the program. Designed to begin slowly, the program would have just ten candidates. If successful, the program would be expanded. Omlie was overwhelmed with applications. More than 1,000 women applied.

The Civil Air Patrol (CAP) was another organization that provided women with flight training. Their stated purpose was to "weld civil airmen and women into a force for national defense by increasing knowledge and skill in every type of aviation activity."[9]

The CAP, while receptive to teenage cadets and women pilots, did not allow the women to fly coastal patrols, although the women trained the men who did fly the patrols.

Ruth Chaney Streeter had earned her private pilot's license in 1941, and her commercial license a year later. She volunteered for the Civil Air Patrol in New Jersey but was told since New Jersey was a coastal patrol area it was too dangerous for a woman. "They told me I could come down and make them coffee if I wanted to," she said. "I looked at the men who were saying this and said to Hell with all of them." Streeter did, however, loan them her airplane for the patrols.

When the WAFS formed in September of 1942, she was one of the first to volunteer. They told her she was 12 years over age and rejected her. Streeter was not one to give up easily. She joined the United States Women Marine Corps Reserve on August 13, 1943 with the rank of major. They put her in charge of aircraft maintenance and air traffic control. By the end of the war she had 19,000 enlisted and 1,000 officers under her command and had risen to the rank of full colonel. She earned the Legion of Merit and was a member of the Women Officers Council.[10]

The International Organization of Women Pilots, the Ninety-Nines, acted as a unified voice for women pilots. Two American women in particular who acted passionately as voices for women in aviation were Nancy Love and Jacqueline Cochran.[11]

Love and Cochran represented two different styles to achieve the same end. In their initial wartime roles, Nancy Love was director of the Women's Auxiliary Ferry Squadron (WAFS) and Jacqueline Cochran, director of the Women's Flying Training Detachment (WFTD). Each woman had her own vision of her role in the war effort, each had very specific political support, and each had her own agenda. They had one common factor, the advancement of women in aviation. Aviation historian Deborah Douglas describes the two women. "Love evoked loyalty and respect from her small group of women pilots. A leader by example, she was an active participant in her cadre of talented flyers. Cochran was, by contrast, a born administrator, skilled in achieving her ends by personal influence and negotiation."[12]

Nancy Harkness Love

Nancy Love was born and raised in Houghton, Michigan. "You know how it is," she said to a newspaper reporter when asked why she got into aviation, "you look at a horse and think, I'd like to ride one. Well, I guess I just looked at that barnstormer's airplane and said to myself, I'd like to fly one."[13] Love learned to fly when she was sixteen, in 1930. A ride in a barnstorming Fleet was her undoing. "I went up for a few flights and I loved it right away." From then on Love knew what she wanted. Fortunately, her parents were indulgent, and she got her private license a month after her first ride. See Fig. 3-1.

Love continued flying in college, receiving her transport rating in 1933 while at Vassar. She pioneered a flying school at the college and augmented her allowance by hopping passengers at the Poughkeepsie Airport. Although Love left Vassar after her sophomore year because of the Depression's effect on her family's finances, she continued to fly. In 1935, she was one of several women hired by the Bureau of Air Commerce to work on its air-marking project. (The others were Blanche Noyes, Louise Thaden, Helen Richey, and Helen Mac-Closkey.) Later Love worked in sales and was a charter pilot for her husband's company, Inter City Aviation, at Boston Airport. Love and her husband were Beechcraft distributors, and went on a California honeymoon in a demonstrator

Fig. 3-1 Nancy Harkness Love

model, one of the first staggerwing biplanes. Love also worked as a test pilot for the Gwinn Air Car Company between 1937 and 1938. Among her duties, she tested the strength of tricycle landing gear by performing as many hard landings as she could.

Nancy Love participated in ferry flights even before the war started. France needed aircraft desperately and in June 1940, she and 32 male pilots were responsible for flying the American airplanes to Canada, where the planes would await shipment to France. The airplanes were to have been used by France, which was under siege by the German Luftwaffe. Unfortunately, however, the German Army occupied France before the planes ever left Canada.[14]

The flights brought Nancy Love into contact with the operations of the Army Air Corps' Air Ferrying Command (known after March 9, 1942 as the Ferrying Division of the Air Transport Command).

Robert Love, Nancy's husband, was a reserve officer in the Air Corps and had been called to Washington to be deputy chief of the ATC. Nancy went with him. She obtained a civilian post with the ATC Ferrying Division Operations office in Baltimore, Maryland, and commuted daily 80 miles roundtrip by airplane. Nancy Love, from her own experience, saw the potential for women ferry pilots and began articulating plans she had in mind for using highly qualified women pilots to ferry military planes. Around this time, Lt. Colonel Robert Olds, then in the Plans Division of the Air Staff, listened to her plan and asked Nancy to list all the women pilots in the United States holding advanced ratings. There were 49 names on this list.

Jacqueline Cochran

At the same time that Nancy Love was working on plans for the Women's Auxiliary Ferry Squadron, Jacqueline Cochran was in the process of arranging a ferry flight for herself to England.

After her meeting with Clayton Knight, the British ATA had invited Cochran to their Montreal base for tests and check flights. Cochran passed the tests (which were extremely demanding) and they gave her a Lockheed Hudson bomber to ferry to Prestwick, Scotland.

Mass protests arose from the male ATA pilots. The male pilots did not want the blame if the Germans shot Cochran's airplane down. Also, the presence of a woman pilot also humiliated them. Prestige and entrenched prejudice were really the issues. Although not always diplomatic, in this instance Cochran worked out a compromise with the ATA. Cochran would pilot the airplane in flight, but her male co-pilot would be responsible for takeoff and landing.

Once in England, Cochran met with Pauline Gower, who was the chief of women pilots with the ATA. Gower repeated Clayton Knight's request and asked Cochran if she could recruit American women pilots to expand the British group. Gower's request got an immediate and positive response from Cochran.

The Royal Air Force (RAF) women, flying since 1940, initially flew only small trainer aircraft. They quickly proved competent and were soon ferrying

Hurricanes, Lancasters, and Mosquitoes from the factories to air bases. The ATA then lifted all restrictions and were actively trying to recruit pilots, both male and female, from other nations.

By this time, Jacqueline Cochran had already hand-picked a group of 22 women to form an American contingent. The women chosen included Helen Richey, the first woman in the United States to fly for a scheduled airline. The women were not only highly skilled but also of incorruptible character. Cochran did not want to risk failure because of the appearance of less than exceptional moral behavior.

Robert Olds (who had been promoted by this time to general) was still interested in Nancy Love's plan to hire a select group of highly qualified women pilots to ease his immediate and increasing shortage of male pilots in the Air Ferry Command.

General Arnold also watched Love's plan taking shape with an interested eye. He liked and respected Jackie Cochran but disagreed with her proposal to Eleanor Roosevelt to create a training corps of women pilots. Like others, he was not convinced that the shortage of personnel required the establishment of all-women divisions. See Fig 3-2.

Arnold was caught off guard when he realized that there was a potential for the simultaneous creation of several groups of women pilots (Cochran's group going overseas, the possibility of a training group in the U.S. and Love's group). The Army Air Force (AAF) leadership had various contingency plans to use selected women but no consensus existed. In the summer of 1942, Arnold ordered General Olds not to do anything about Love's proposal until Jacqueline Cochran returned from England. Any action by Olds to create a women's flying corps headed by Love would have put Arnold in a paradoxical situation. He had promised Cochran, under pressure from the White House, that she would direct any women's program instituted by the AAF. Arnold knew that Olds disagreed with Cochran's plan for a training program, and given the first opportunity would give Love the green light with her plan.[15]

Nancy Love's proposal assumed select women with advanced ratings would follow the same path already established for the men of the Ferry Division: they would be hired as civilians, given a 90-day trial period and then commissioned into the AAF. There was a hitch, however. Legislation then did not allow for flying officers and flight pay for women pilots and, therefore, they could not be legally commissioned in the AAF.

Gambling that legislation would eventually pass to permit women pilots in the military, especially once women demonstrated their abilities, Love proposed that the ATC go ahead and hire the women as civilians. Love also made it clear that the women's status was temporary and that she expected the ATC to make every effort to get the women commissioned.

Love recognized that her pilots would be examined carefully so she stiffened the requirements for women candidates. They had to have a 200-horsepower rating and two letters of recommendation. Love, like Cochran wanted women with outstanding personal conduct. Love also proposed that the women should earn $250 per month, $130 less than the male civilian pilots received. Her reason for this was that the women, she said, would be flying only small training and liaison-type aircraft.[16]

According to Jacqueline Cochran, however, Arnold knew nothing of the WAFS plan and after seeing it in July said there did not appear to be any official record of authorization. Nevertheless he did not attempt to kill it.

Pressure was mounting for a solution to the problem. Six months after Pearl Harbor, the pilot demand continued to outstrip the supply. Even after lowering the draft age, and relaxing many preflight school restrictions, the United States could not supply the demand. Besides supplying pilots for combat assignments, there was a need to fill more non-combat flying jobs. These included testing new aircraft, ferrying new airplanes from factories to air bases, delivering supplies, medicines, and equipment and flying target, searchlight practice, and VIP flights.

Fig. 3-2 Jackie Cochran with General Arnold

After a long battle to bring women into military aviation, Nancy Love was about to succeed. The Secretary of War, Henry L. Stimson instructed Love to form an experimental group of experienced women pilots to ferry aircraft for the Air Transport Command. Love sent telegrams to 200 women each of whom had at least 500 hours of multi-engine time. She received 100 replies from women eager to join her squadron.

General Arnold, at the direction of Stimson, was to announce the formation of the WAFS with Nancy Love as their director on September 10, 1942. Two things happened at this time. On September 5, Cochran was still in England but preparing to depart on a flight for the United States. General S.H. Frank, of the Eighth Air Force, had Cochran returned from the airport, ostensibly on an urgent matter. Cochran waited three days to find out what Frank wanted. The matter turned out to be a thank-you dinner. She was not back in the United States on September 10.

The second thing was the unexplained absence of General Arnold on September 10th. Secretary of War, Henry L. Stimson, made the announcement. It was obvious that Arnold did not want to bruise his relationship with the White House.

So that Jacqueline Cochran would not feel upstaged, on September 14, 1942, she was authorized to take charge of the women's pilot training. Her job was to train at least 500 women pilots with limited air time and turn them into trained ferry pilots.

By October 22, after passing the same physical exam and flight test given to men, 27 women were selected for Love's ATC. The women went on to learn Army Air Force flight procedures, navigation, meteorology and codes - the same training as the men received.

Most of the ATC, however, still needed convincing that women could really fly. To test the women's metal, the ATC subjected them to every test and ground school class imaginable, even when the women had more experience than their military instructors. The women passed all the tests with high marks and in December 1942, ATC General George said: "If they can fly four-engined bombers safely after proper periods of training and preliminary work, I see no reason now why they may not get the chance."[17]

Women's Auxiliary Ferry Squadron (WAFS)

Nancy Love's original twenty-seven WAFS were an elite corps. Among the most experienced women pilots in the nation, they were articulate, bright, and enthusiastic. They exuded an aura of good humor and self-confidence. Many of the WAFS were as skilled at flying as Cochran, and they watched her maneuvering with a jaundiced eye. It would take more than flight experience for Cochran to earn their respect and the right to leadership among them.

Cornelia Fort[18] who was in the air over Pearl Harbor with a student and narrowly missed getting shot down, was the first woman accepted. She had 845

Fig. 3-3 Evelyn Sharp

hours. Fort was later to be the first of 38 women to die in the service of the WAFS/WASPs. Evelyn Sharp, who had 2,950 hours and who was the youngest woman transport rated pilot at the time, prior to being accepted into the WAFS. See Fig. 3-3.

Their training would consist of one hundred hours of flight training and an eighty-hour ground school course. While in training they received $150 per month, out of which they paid $.75 a night for barracks, and $.35 to $.50 a day for meals. Away from base they received an additional six dollars per day to cover travel expenses.

Unlike the WACs and the WAVES of the Army and Navy, the WAFS did not, upon graduation, receive a military status. (Although Love continually pressed the issue, the needed legislation was never passed.) They served on a civil service basis, working for the Army, and were subject to the Articles of War. They were treated like officers in the Army. For a violation of the Uniform Code of Military Justice they were liable to court-martial the same as an officer in the armed forces.

Love's personal duties involved administration of six ferrying squadrons and planning the operational and training procedures. In addition she had a chance at the fun part of the job. Love went through transition on each type of aircraft. She also ferried at least one of each before that type was released for training and ferrying. "The first exciting plane I flew," she said, "was the P-51. I continued through the B-25, a B-26, an A-20, a P-39 and so on to my goal of a B-17. I had always wanted to fly a B-17 and wondered if I could handle an airplane of that size."[19]

Love and Betty Gillies, her second-in-command went into B-17 transition school in August 1943, and in two weeks they were ferrying the bombers across the country.

By this time, the build-up of aircraft for Europe was in full swing. Colonel William Tunner had orders to fly two-hundred B-17s to England. Most of Tunner's recently assigned male pilots were from the training detachment and were reluctant to fly their first operational cross-country mission across the Atlantic Ocean. Tunner wanted to prove that the mission was no different from any other ferry mission. He asked Nancy Love and Betty Gillies if they would ferry a B-17 to England. (The old ploy, "if a girl could do it..." was still alive and well in the midst of war.) Love and Gillies agreed. Two women would finally get to fly "across the pond," without male supervision.

Tunner knew the flight would cause controversy so, five days before the scheduled flight, he dispatched a wire to all ferry commands saying that there would be no publicity around this flight. Without fanfare, the two women started for Prestwick, Scotland, with a crew of three enlisted men and Tunner's personal navigator, a male lieutenant. Weather held them overnight at Goose Bay, Labrador, and their dream began to wash away in the storm. General Hap Arnold caught on to the mission through a harmless bit of military courtesy. When the British Air Minister showed him a telegram announcing the planned arrival of the two women, Arnold immediately grounded Love and Gillies. "Betty and I, raging and frustrated, were sent ignominiously home, while a male replaced us," said Love.[19]

Nancy Love immediately suspected Cochran of complicity in grounding the flight and remained convinced of this for a long time. The speculation around this incident was not the attitude of the women being in danger making the flight, but the promise on Arnold's part that Jackie Cochran would get favored treatment (as a friend of the Roosevelts). The success of this flight would have elevated Love into a high profile that would have put Cochran into a secondary position of importance, and unacceptable to Cochran personally.

Women Airforce Service Pilots (WASP)

The original twenty-seven women selected for the WAFS were supplemented by women trained under Cochran's WFTD and by August 1943, over 100 women were ferrying aircraft for the WAFS. Although originally designated to fly liaison aircraft, their level of enthusiasm and experience quickly earned the WAFS seats in pursuit aircraft.

On August 5, 1943 the two women pilot programs merged, and Cochran's Women's Flying Training Detachment and Love's Women's Auxiliary Ferry Squadron, became the Women Airforce Service Pilots program (WASPs). This new organization provided a necessary and vital link to the war effort, and a significant advancement in the growing list of American women in aviation. Nancy Love reported to William Tunner, commander of the Ferry Division. Cochran became director, reporting to Hap Arnold.

The women who flocked to Avenger Field in the blistering heat at Sweetwater, Texas, for WASP training came from many life experiences. Marion Florsheim (of the shoe fortune) arrived in Houston with 17 steamer trunks and an escort of Afghan hounds; Katherine Landry had been a riveter on a Ford assembly line. Helen Dettweiler had been a golf champion, and Marie Shale of Reno was a former blackjack dealer. There were Ivy Leaguers, shipfitters, writers and an Olympic swimmer.[20] Their common bonds were their enthusiasm and love for flying. Some had washed and repaired airplanes to get the money to take flying lessons, at a time when a women's annual salary was less than $800.

Out of a volunteer pool of over 25,000 women, 1,800 were selected and 1,074 women earned their silver wings. They flew every type aircraft in the American war arsenal from 1942 to 1944.

America's attention was centered on the war overseas and most of the general public was not aware of the WASP effort. Civilians often mistook WASPs in uniform for airline stewardesses and even military personnel found them oddities. A typical reaction from many male pilots upon seeing a WASP is best expressed by Jean Landis, a WASP. "Reactions to women climbing out of a P-51 were varied, mostly startled. I flew into a field that was off limits, but the weather was bad and I had a mechanical problem so I called for permission to land. I kept radioing, 'P-51 ready to land; awaiting final instructions.' It was somewhat garbled and they kept asking me to call in repeatedly. Finally they said, 'Waggle your wings if you receive.' Then, 'Lady, the only thing we see up there is a P-51.' They couldn't believe it - they were looking for a Piper Cub or something. They were darlings. By the time I had taxied up to the line, following the little *Follow Me* truck, there were lots of guys around to see what kind of woman was flying a P-51. They had never heard of the WASPs."[22]

In some instances their uniform of jacket and slacks was a hinderance. They were sometimes excluded from nightclubs and restaurants where women in slacks were inappropriate. In Chicago a train was held for several WASPs so they could get back to Romulus Army Air Force Base quickly. When the women arrived in slacks, with parachutes slung on their shoulders, the officials refused to let them on the train. They claimed the women must have been women that male pilots had picked up.[23]

An anonymous author also stated in a pictorial article in *Look* magazine dated January 9, 1943, "The U.S. has found it hard to believe that women make

Fig. 3-4 A group of WAFS

good pilots. Like women doctors, they have grown in number and amassed their impressive records of achievements through sheer feminine doggedness."

PROFILES IN COURAGE

Kay Brick

Katherine Adams Menges Brick was born August 8, 1910 in Dixmont, Maine. She had her first airplane ride on Christmas day in 1934. Kay went on to graduate from Sargent College with a B.S. in education and then earned an M.A. in psychology from New York University. She did further graduate work at Columbia University and then decided to teach at Lascell Jr. College, in Auburn MA, specializing in dance and aquatics. From there she moved on to become director of physical education at Staten Island Academy. Teaneck High School in New Jersey was offering adult education courses in engineering and aerodynamics and Kay signed up. Soon she realized that aviation held the key to the future and went to Teterboro Airport for flight lessons. Her flight instructor was a good-looking young man named Frank Reeve Brick whom she would marry.

Kay earned her pilot's license and bought a used Aeronca Chief. Just before the war broke out the Civil Air Patrol was actively recruiting women and Kay enlisted as a lieutenant based at Teterboro Airport.

When Kay heard about the formation of the British Ferry command, a Women's Auxiliary Air Service in England formed to ferry war planes she went down to enlist. Kay had been accumulating several hundred hours in Frank Brick's 450 horsepower Fokker Universal in preparation for such a group. She found to her disappointment that she needed much more time in the air and several ratings. She went ahead and began building her time and perfecting her flying skills. She flew endless hours at 800 feet perfecting double precision entry and exit spins to gain her commercial license. Then when she felt she had the qualifications, she went down to enlist, again. It was then she found that the government was no longer permitting American women to go overseas to join the British Air Transport Command. She then went to the WAFS and just as she found out she qualified, Jacqueline Cochran told her about the WASP program. Kay Brick immediately signed up. See Fig. 3-5.

Kay soon found herself in Sweetwater, Texas, at Avenger Field and graduated in the class 43-3 as an advanced Ferry Command Pilot with single and multi-engine ratings.

She went on to New Castle Army Air Base and eventually on to Stewart Air Force Base, in Georgia, where she graduated from the radio-controlled aircraft school. Under top-secret conditions, Kay flew a PQ-8, a radio-controlled target plane. She sat in the tiny aircraft while a second WASP used radio controls in another ship, to fly the plane. Kay would sit "captive" as her tiny plane zoomed and dived and maneuvered wildly through the skies. (In an emergency she could override the controls and take control of the aircraft.)

Kay Brick had other exciting rides as a WASP. She checked out in what the fighter pilots called 7-tons of Hell, the P-47 "Thunderbolt," affectionately knows as "The Jug." While at Biggs AFB, in El Paso, Texas, she served as squadron commander. Some of her missions included high and low target towing, tracking, laying smoke screens, and "buzzing" to check camouflage of anti-aircraft gun emplacements, flying formations through chaff, split beam searchlight missions and low altitude night missions and ferrying operations.

After the war, she married Frank Reeve Brick and they had one daughter, "Bunny." The school held Kay's pre-war teaching job for a year after the war ended, but Kay could not go back inside. Someone referred her to L. Bambergers & Co. which had just started an aviation department. For two years she sold airplanes, gave instruction and promoted aviation. Kay joined the Aviation/Space Writers Association, which allowed her many more experiences in military aircraft and opportunities to do such stories as on-board aircraft carrier missions. She sold and ferried aircraft around the United States and Canada and made several trips to Alaska, and one to Australia. She entered air races and flew one race from New York to Miami, and then flew her BT-13 to Cuba.

In post-war America, those WASPs who had earned their silver wings received commissions in the Air Force Reserve. A year and a half after Kay received her commission, someone discovered her time warranted a higher commission. This update of her records led them to notice that she had a child under 18. Under the Armed Forces Act of 1952, (and para. 10e, AFR 35-6, 3 Feb. 53 as amended) any reservist with a child was released from assignment and discharged under honorable conditions..." Kay Brick replied to officials that "My mother lives with me and cares for my daughter; my husband supports us." The official reply to Kay was, "Give up your child for adoption...or give up your commission." The solution to the government's problem was easy. Kay resigned her commission.

Today Kay is encouraged. "Things have changed now. Military women now marry, have children and often get posted with their husbands at the same base. Some younger women are amazed that such a rule ever existed."

Kay Brick also became active in other organizations. As a member of the Ninety-Nines, she rose to become its president (1950-1951, and was on their executive board (1951-1954). She also served as executive director on the Powder Puff Derby (All Woman Transcontinental Air Race) for 13 years, and served on the board of directors for 25 race years. She was also secretary on the board of directors of the National Pilots Association, (1956-1960) and 1968-1970) and held the post of secretary on the board of directors to the P-47 Thunderbolt Pilots Association (1970-1974, and 1978-1980). She remained an officer in the Civil Air Patrol (captain), and received an appointment to the FAA's Women's Advisory Committee on Aviation (1968-1971). Her other post with the FAA included Safety Counselor, (1972-1976).

Kay also was the initiator/chairperson of the committee that obtained a "votive" model of Amelia Earhart's Lockheed "Vega," the plane Earhart flew solo across the Atlantic.

Kay Brick has won many awards during her aviation career: the Lady Hay Drumond-Hay Award from the Women's International Association of Aeronautics (1946); the FAA's Certificate of Commendation (1966); Amelia Earhart Medals (1949, 1960, 1976); the Paul Tissandier Diploma from the Federation Aeronautique (1973); and the International World Body Award in 1973. She was inducted into the Aviation Hall of Fame of New Jersey in 1978.

Today Kay can take credit for flying more than 60 different aircraft more than 6,000 hours. In September 1991, she renewed her pilot's license, making it fifty years that she has had an active license. She remains a member of the Aviation/Space Writers Association, the OX-5 Pioneers; Silver Wings Fraternity; Altrusa International; and the Air Force Association. Kay gave passing thought to becoming involved in the space program but realizes in 1992, her "timing may be off."[24]

Among her other accomplishments, Kay is the author of *Thirty Sky Blue Years - Ninety-Nines History,* and she edited the *Powder Puff Derby Commemorative Album, Powder Puff Derby - The Record,* and the *Powder Puff Derby Commemorative Update.* In 1992 Kay became a member of the United Flying Octenegerians (UFOs) and was the youngest of the five oldest, currently

Fig. 3-5 Kay Brick

licensed women pilots. She currently serves as the vice president of the Silver Wings Fraternity.

Kay Brick has seen many advances in aviation over the years. Her greatest pride is knowing that "WASP accomplishment formed the building blocks of the present. We know the future is predicated on the past - more than ever since the doors for women have opened wide. Pick your goal," she says, "and do it! An aviation career for a woman today presents such an exciting world - truly around the world, above and beyond."

Marjorie Gray

For Marjorie Gray, flying changed her whole life for the better. She learned to fly in 1938, in a 40 horsepower Taylorcraft, at the North Jersey airport in Franklin Lakes, New Jersey. She was fortunate to have the support of her friends and family during and after her initial flight training. Marjorie earned her commercial certificate in 1942. When World War II broke out, Marjorie volunteered for the first class of women to undergo Army Air Force flight training. "We were the guinea pigs," she said. She went on to serve 25 months in the Women's Auxiliary Ferry Squadron with the Air Transport Command. During her enlistment she flew 19 different types of military and civilian aircraft from

Fig. 3-6 Marjorie Gray

the Piper L-4 Cub to the heavy B-25 bomber and C-47 (the military version of the Douglas DC-3), logging more than 750 hours. See Fig. 3-6.

When the WASP disbanded in December of 1944, Marjorie Gray went on to open her own fixed base flight operation (FBO) business in Teterboro, New Jersey. She was one of the first woman-owned FBO's in the United States when she established Marjorie M. Gray Inc. Aero Services, and she operated the facility at Teterboro, New Jersey from 1946 through 1951.

In those days, operating an FBO was considered by many "men's work." That did not deter Gray. At first she had three airplanes but no hangar, and had to buy gas at retail prices. She worked hard and within a year had a 60 by 80 foot hangar, and was buying gas in bulk at discount prices. Marjorie Gray worked her FBO for twenty years before selling it and joining Curtiss Wright as an aviation technical writer. Later she worked as an associate editor for *Flying* magazine.

In 1956, Marjorie Gray received the Lady Hay Drumond-Hay Award for "Outstanding Achievement in Aviation."

Gray also received a commission in the Air Force Reserve. She retired from the Air Force as a lieutenant colonel and today enjoys attending meetings of the Ninety-Nines and watching the progress women have made in aviation.

In 1992, Marjorie Gray took her place among the other aviation greats like Amelia Earhart and Kay Brick when she was inducted into the Aviation Hall of Fame of New Jersey.

Barbara Erickson London

Barbara Erickson transferred out of her college home economics courses when a girlfriend talked her into applying for the civilian pilot training course being offered at the University of Washington. In 1939, the program was one of many sponsored by the government at various colleges to increase the pool of potential pilots for the coming war. One woman was allowed to participate for every ten men accepted into the program and although her girlfriend flunked the flight physical, Barbara passed.

Better suited for flying than home-economics, Barbara quickly excelled in the flight program. She advanced through her ratings and became a flight instructor for the government program within twelve months. She graduated in 1942 and won the Northwest regional competition for the outstanding Civilian Pilot Training Program that same year. "Learning to fly changed my entire life," recalls Barbara.

Like all Americans eager to lend their talents to the war effort, Barbara considered going to England to fly with the ATA, but when the United States entered the war in 1942 she joined the WAFS. "The Army Air Corps opened up the first Women's Auxiliary Ferry Squadron at Wilmington, Delaware. They sent letters to about 100 women in the United States who had commercial licenses with 500 to 1,000 flying hours and a 240 horsepower rating. I went to Wilmington, checked out, passed the physical and became a ferry pilot in September 1942."

At the age of 23, Barbara was made a squadron commander over 80 women pilots of the 6th Ferry Group in Long Beach, CA. Besides ferrying airplanes, her duties as a commanding officer were to maintain the discipline and morale of the women and act as a liaison with operations. Barbara's unit was the only female squadron of the five military units stationed at Long Beach and her women flew every aircraft built, from basic trainers to P-51s. California was the center of aircraft manufacturing with companies like; Northrop Douglas and Lockheed. The ferry pilots had their work cut out for them. "We picked up planes at the factories and flew them to embarkation and modification centers all over the U.S. We flew P-38s and P-51s to Newark for the European war and P-47s, P-39s and B-29s to Alameda to put on a boat to go to the Pacific." [25] See Fig. 3-7.

Arrival of women at the factories and delivery sites drew reactions ranging from amazement to disbelief, as women had never before been seen flying the military's machines. "I was delivering an A-26 to a training post for pilots and called the tower to tell them I was on downwind. They said they didn't have me in sight and instructed me to follow the A-26 on downwind. I told them that I didn't have the other airplane in sight but that I was turning base. They replied that they still didn't have me in sight and to follow the A-26 on base. I landed and taxied underneath the tower and they would not believe that there was a woman in that airplane!" [26]

The WASP's schedules were never very definite and at times the women had to be ready to fly at a moment's notice. Often they would fly their own

Fig. 3-7 WASP Barbara Erickson boarding a P-47 fighter

airplanes out to a factory to pick up an Army plane that was ready for delivery and would reposition themselves using commercial transportation. "When flight crews discovered that I was a WASP they would ask me all kinds of questions about the equipment I was qualified to fly. Once a DC-3 co-pilot asked me if I was checked out on that airplane and I told him that I was. 'Jeepers,' he said, 'I've been riding from the right seat of this DC-3 for ten years and I still haven't checked out as Captain.' When he asked how long it had taken Barbara to checkout on the DC-3 she replied, "Three or four hours!"[27]

Aircraft production soared in 1943 and the rotation rosters kept the women flying constantly. Barbara drew a rotation that had her complete a record ferrying mission of four 2,000 mile trips across the U.S. in a single five day period. She left Long Beach on the morning of July 29, 1943 and delivered a P-51 Mustang to Evansville, Indiana before dark. The following day she flew a P-47 back from Evansville to San Pedro and rested overnight on the way. On August 2, she ferried a C-47 from Long Beach to Fort Wayne, Indiana and completed administrative duties at Ferry Division Headquarters in Cincinnati until August 6th when she returned to California in a P-47 Thunderbolt. The deliveries would normally have taken weeks to accomplish and Barbara had managed it in a matter of days. She covered more than 8,000 miles in about 40 hours. [28]

Erickson was also one of the most highly qualified pilots in the WASPS. At one point she had checked out in 36 aircraft. Most WAFS/WASP ferry pilots were current in 10-12 airplanes at a time. [29]

The ferrying feat was a fine example of the unprecedented effort and the distinguished service of which the WASPs were capable. Barbara was commended for her outstanding ferry service and was decorated with the only air medal ever presented to a female pilot during World War II. The air medal for meritorious achievement was pinned to Barbara's lapel by General Henry "Hap" Arnold, commanding general of the U.S. Army Air Forces, at Avenger Field in Sweetwater, Texas, during the graduation ceremonies of class 44-W-2 on March 11, 1944. See Fig. 3-8.

As the war drew to a close the men overseas returned home to flying positions previously held by the women. "We were simply replaced. It was heartbreaking for us because we left before the war was over and before our jobs were done. It was particularly hard for me because the morning I left Long Beach there were sixty-one P-51s sitting in the middle of the field that my squadron could have delivered that day. Instead our command had to borrow pilots from other units to get the airplanes delivered."

"We flew literally thousands of flights and our safety record was incredible. We lost five gals in my squadron, mostly due to training accidents. I was very proud of the women I flew with during the war years. They did a job every bit as well as any pilot in this country." [30]

When the WASPs were disbanded in 1944, Barbara married a military pilot, Jack London. In 1947 she received a commission as a major in the Air Force Reserves and completed twenty years of military service before she retired. As a civilian, Barbara became the executive director of the "Powder Puff Derby" and led the successful All-Womens Transcontinental Air Race until the mid 1960s when she went into aircraft sales.

For the past two decades Barbara has been the owner and operator of Barney Frazier Aircraft and is still firmly rooted in Long Beach where she has remained ever since the war. She raised a family there and even inspired her daughters to fly. One daughter has flown as her company pilot for more than ten years and the other became the first female airline pilot at Western Airlines in 1976. Today she is an active member of the Airport Advisory Commission and strives to bridge the gap between the community, pilots and airport authorities on sensitive issues that threaten the very existence of our general aviation industry. For her years of dedication and involvement at the Long Beach Airport, the city fathers recently bid her a fitting tribute in celebration of her extraordinary aviation commitment and service.

Of Barbara's most recent honors and awards, she was extended the elite privilege of attending the annual "Gathering of Eagles" in 1991. The event highlights prominent aviation pioneers like famed "Eagles" Chuck Yeager, Neil

Fig. 3-8 Barbara Erickson receiving the Air Medal

Armstrong and Buzz Aldrin who, with their vision and inspiration, have helped shape aviation's heritage and history. It is a unique opportunity for aviation trailblazers like Barbara to share personal accounts of their history-making endeavors with the graduating class of the United States Air Force Air Command and Staff College.

While it is true that Barbara's impressive aviation career has spanned more than 50 years and that she has lived at the very threshold of aviation's growth and development, she recounts the military ferry days to be among the greatest years of her life. "The women of my era were so lucky to be flying during that period in our history. To be twenty-one and flying every piece of equipment being made was a pilot's dream come true and an opportunity that women will never have again."

The End of the WASPs

In December 1944, the Army issued a CONFIDENTIAL report on the WASPs.[31] The report, written by a forever anonymous male in effect doomed the WASPs with half-truths, innuendo and lies. It is obvious from the report that the author was vindictively jealous and probably a flight school washout.

"A single magnificent factor has come to light which, admitted by some WASPS, denied by a few others, has yet been recognized throughout the flying world.

Women, it seems, are essentially unmechanical; theirs is an innate lack of mechanical aptitude which makes flying, as practiced by the majority, of an unsafe caliber because it is so mechanical. In the opinion of one of the heads of Air Training at this (unnamed) station, only five percent of women flyers throughout the world are possessed of an inherent flying ability, present with few exceptions in male pilots."

In the midst of a world war, how anyone could gather world-wide statistics should make one wonder about the credibility of the report. It will, however, remain a mystery.

Until the fall of 1943, the Ferry Division of the Air Transport Command accepted men with 300 hours total flying time, 200 of which had to be in ships of 200 horsepower or better. The original WAFS were required to have 500 hours, and entered the service on the same status as the men.

The report continued, *Most women will assimilate the knowledge that certain actions must be taken to accomplish certain ends, but the connective aspects of mechanical accomplishments they do not know make necessary transition for this upgrading greater than the time usually required for similar upgrading of male pilots. Ground Trainer time was also required, and this section of training also testifies to the basic inadequacy of women in flying. It seems that the WASP find difficulty in performing two or three tasks at the same time, at least in the Ground Trainer. The daily grades earned by the WASP in this section are, in the main, only average or below average. This might have been because of a simple failure in concentration or merely the attachment of small importance to this work."*

The only variation in training between the men and the women was according to Nadine Nagle, a WASP, one of quantity in one area. "The men had more aerobatic training to prepare them for dogfights. The women had more navigational training for cross-country flying." [32]

The report also ignored the dangerous work the women did. Officially they were known as Engineering Test Pilots. In that capacity their job was to test fly any airplane that had a major overhaul before the men flew them. The women were expendable, the men, in a non-combat environment were not.

WASPs trained male bombardiers with live bombs. On one flight a live bomb did not drop free and jammed in the open bomb bay. The WASP pilot was determined to save the aircraft. After the male crew chief tried unsuccessfully to kick the bomb loose, the WASP pilot landed the plane with the loose bomb hanging in the open bomb bay. She remained cool and collected and did not think anything of the incident.

The historical officer's report ignored completely the WASP record of accomplishments. They flew 60 million miles, delivered 12,000 aircraft and their outstanding safety record compared favorably with men doing the same work.

This record was achieved inspite of the fact that when weather grounded the male ferry pilots, the WASP were often ordered to fly. Again, they were expendable.

The report continued, *"One of the largest detrimental factors in the use of these women pilots was the fact that they retained a civilian status and for this reason can disregard military discipline and regulations in the faces of those men who were over them. It was not unusual for a WASP to report late for an assignment, nor could much be done in the way of punitive measures for such action; resentment of being required to take certain ground school subjects was often manifested by a failure to report to the class at all. While not the general rule, there were cases of direct disobedience to orders, and this was the focal point of much of the resentment and strain between WASPs and male pilots."*

This claim completely overlooked the fact that women were treated like officers and like all other civil service employees in the Army were subject to the articles of war and the discipline associated with it. They were also subject to military discipline and court martials for offenses.

Nagle would probably challenge this claim also. "They were grooming us to be officers and we were treated accordingly. Discipline was very strict. We were not supposed to associate with enlisted men."[33] Nagle was referring to the 460 women who enrolled in a three week Basic Military Training Course for Officers, at the School of Applied Tactics, in Orlando, Florida. All the women enrolled passed the course.

Ruth Roberts, another WASP, wrote a letter to the editor of *Time*. It expressed the feelings of all the WASPs.

"We have been ordered to attend Officer Training at Orlando, though no woman in this service during the twenty months of its organization has ever asked to be commissioned. Women flyers volunteered to do the work of men needed overseas without the prestige of rank at a time when men were not volunteering in sufficient numbers to cover ferrying assignments."

In contrast, the report continued, *"One opinion advanced has been that the girls are overly anxious for publicity and public acclaim; in some cases, the writer has been told, this became almost an obsession."*

The women were also subjected to demerits for infractions, just like the men, but there was a significant difference in their application. The men were allowed to "walk off" the demerits. The women were not allowed to walk them off and if they accumulated 70 they were washed out of the service with an immediate and permanent release. Also, if the men could not perform a particular type of duty they were assigned to other duties. The women again were washed out.[34]

One WASP trainee got lost on a cross-country flight. She was flying an AT-6 and the rule was if you had to put it down in an emergency, make a wheels-up landing. The trainee found a small field and decided she could land without damaging the aircraft. She landed without putting a scratch on the plane. She was also washed out of the program just two weeks before she was to have graduated. She had broken the Army's rule of landing wheels-up.[35]

On May 8, 1944 *Time* Magazine ran an article on the WASPs. The WASPs were quick with a rebuttal. Elizabeth Hartz wrote the editors pointing out errors and unjust implications of the article.

We are frankly tired of the limelight - from "Ladies Courageous" to congressional accusations. We are a group of girls; your sisters, wives, and daughters who have left our jobs and homes because we were told there was a job for us to do with winning the war. It's not been easy; it's not been all fun. We decided to join, most of us, a year and a half ago when many present mud-slingers were still evading the draft and earning fat war-boom salaries. Some of us didn't make the grade; some have been killed in "the line of duty."

We fly without protection in the form of hospitalization and insurance. But we are young; we don't think of the risks.

As for the facts, those I leave to the ones who quote them readily. This is merely a plea. Cut out the publicity, good or bad. Let us do our job as civilians or as Army personnel. It is the job that counts. We will serve the Army Air Forces as long as we are needed to help with winning the war. And when we are no longer needed, we will quietly and thankfully return to our homes with the knowledge that, to the best of our ability, we have done a job that was worth doing."

Evelyn Sharp died on April 3, 1944 flying a P-38. Her WAFS comrades found it necessary to take up a collection to have her body shipped home. Her mother was not allowed to display the Gold Star, signifying the loss of a son or daughter to the war.

For the men and women in the military branches who were killed in combat, their next of kin received a $10,000 insurance policy, and a casket with the remains shipped home, with expenses paid by the government. They also received full military honors. For the WAFS/WASP it was like the government was saying, "Thanks, here's your daughter back."[36]

The historical officer's report even accused some WASPs of grand larceny. *There have been cases of unforgivable neglect in the execution of missions. A specific instance was that of a WASP delivering an aircraft to a northern point and being talked into selling the ship to the personnel of the base without approval from her home station. It is inconceivable that a military pilot would commit such a breach.*

Most of the WASPS seem to realize that their adaptability and proficiency will decrease when certain types of aircraft are approached. Their physical limitations will deter them from any desire to operate in such ships as the B-24, B-17, C-54 and other in the same class.

At the time of this statement at least seventeen WASPs had checked out in the B-17 bomber. Nancy Love also checked out in and ferried military C-54's (a four engine DC-4 civilian airliner) in the U.S. "This was fun," she recalled, as there were always large numbers of hopeful military personnel thumbing their way on leave or change of station. Love and her WASP co-pilot filled the plane on every trip. She had in effect became an "airline pilot," albeit temporary and three decades ahead of her time.

As far as women having physical limitations, the glider tow program is a typical example that negates the allegation. The glider tow program was experimental. The male officers said the women were weak and too erratic for such an assignment. The WASPs successfully flew not only this but many other experimental programs at the request of General Arnold. The reasoning was if the experiment failed, it could be canceled with no male pilot losses. If it succeeded then it could be made operational and handed off to the male pilots.

The glider program was a success and the women demonstrated that they could tow not one, but two gliders with 15 men and a jeep aboard. The men in the gliders hauled through the black night at barely a hundred feet above the trees expressed a definite preference for a WASP pilot. There had been too many cases of male pilots falling asleep and driving the wooden ships right into the ground.

The women did not ask any favors from the men and there are many WASPs who will testify they never allowed a man to carry their parachutes to the flight line, even though there were some men who thought that was one of *their* jobs.

WASP Byrd Howell Granger related that on one check ride, her instructor was taking her through spin recovery. "He was very sloppy on the controls," she said, "and heavy on the rudder." She had more than 300 hours and had done some instructing, but deferred to her instructor.

On the first spin, he pulled the nose up too high. "I said, hey, we're liable to go into a secondary spin.

"The instructor said what did I know, after all, he was the instructor." Again she deferred to him.

On the second spin recovery he pulled to nose too high and the aircraft went over on its back and slid into a secondary spin (where the tail is forward of the engine in a vertical plane).

"He froze at the controls and I was pumped full of adrenalin. I was able to pull up but much too close to the ground. He didn't say a word or touch the controls. I flew the airplane back to the base. He got out of the plane and quit on the spot. It was for the best, eventually he would have killed someone. . .. But for every bad male pilot we encountered there were a dozen good men."

Then there was the B-26 bomber, called by many "The Widow- Maker." Male pilots were so in fear of losing their lives in this ship that a group of them took off their wings and refused to fly it. When Arnold heard this, he was furious. He ordered a WASP class to report to B-26 transition school at Harlingen, Texas. Before the end of the day every WASP had checked out in the infamous plane. The men had stood by all day watching. The next day, after Arnold had received the results, he issued orders that any man refusing to fly the bomber would be reassigned to the infantry. The men pinned on their wings and flew.[37]

"*The question of conduct can be gauged only by hearsay evidence;* said the report. *The WASP record as available at this station is spotless in this regard*".

Even with no evidence presented in the report to support misconduct the author still raised the question. The Historical Officer also raised the question whether or not the American public got its money's worth out of the WASP program:

"Let us consider what are these women whose organization is soon to be deactivated to do after they are released from their present duties. No one will argue that there remains no responsibility on the part of the WASP to flying itself, to the training that they have had, and to the taxpayers who have paid and will pay for that training."

In August 1943, the Truman Civil Service Committee alleged that it cost over $22,000 per student to train flyers under Cochran's WASP program. The actual cost of a WASP's training was $12,150.70, a figure that included plane depreciation and maintenance, medical examinations, functional flight clothing and salaries. The amount was comparable to that paid to train male flight cadets.[38]

The report did not mention this or the fact that the women earned about $7-12 a month take home pay and at the same time, the male civilian organization Aviation Enterprises was paid $6,540 for each student.

The report continued: *They have been flying types of aircraft that are available only in the military service today; types that cannot be found outside even if time on them was not beyond the reach of individual pocketbooks. Are they to go back to flying the small Cubs and Aeroncas and Taylorcrafts for purely individual satisfaction and thrill? There is a very definite way in which their training and experience may be partially or fully used. It seems to have been proven that there is now no place in the military machine for these feminine flyers, but there is in them a wealth of experience and basic flying knowledge; that which they have gained in the last two years should not be lost. There is no conceivable reason for expecting that they will find it possible to gain employment as factory-employed test pilots on heavy aircraft after their deactivation. If, however, each of these women were to train or begin the training of as many civilians as possible* (presumably males) *in flying, there is not one voice that would consider the expenditure on the WASPS as anything except well worthwhile. For the most part these women are undecided as to exactly what they shall do.*

Some few months ago, when consideration was being given to the advisability of giving them a military status, few might have refused the opportunity, however, and elected to cease their activities. Presently those WASPs who remain seem to feel that they would accept military status if a definite need could be proven. There can be no way to judge the sincerity in this, present conditions being as they are."

Statements made during and after the WASP tenure clearly reveal that the Historical Officer did only superficial research and exercised complete bias against the WASPs. He did so from the safety of anonymity and position. Because of its "Confidential" stamp, he was sure that Cochran or the WASPs would never see the report. It is also fairly certain that Cochran never saw this report. Knowing her personality, she would not have stood still while her beloved WASPs were accused of everything from mechanical ineptitude to grand larceny. It was also clear that he had no idea of their wide-spread use (at 130 air bases) and accomplishments.

He also neglected to mention that the WASP, were given some of the dirtiest jobs too. They had to ferry war-weary planes to the junk yards. Planes that should have gone by rail. Planes that had bald tires that would blow on landing and planes that had used low octane fuel which often caused engine failures.

The role of the WASP was so critical to the war effort that they carried a #2 priority on civilian airliner flights. Only the president of the United States could bump them from a flight.

Postscript

In spite of the confidential report, the WASP's record and courage spurred General Arnold to state at the last graduation ceremony of WASP pilots: "*It is on the record that women can fly as well as men. In the case of another national emergency, we will not again look upon a woman's flying organization as experimental.*" He added, "*If there was ever any doubt that women could become skilled pilots the WASP have dispelled that doubt.*"[39]

The women pilots of Britain and Russia, who were doing ferrying and transport work, proved sex was no barrier to flying ability long before the women in the United States were accepted by the Air Transport Command. General Arnold knew about the British women as early as 1939. The Russians, however, were more secretive about their "Night Witches."

Although American military services and other Allied countries did not allow women to fly in combat, the Russian government had different thoughts. The Russians can claim the first woman pilot in history to shoot down an enemy bomber. Valeria Ivanovna Khomyakova shot down a Junkers JU-88 in late 1942 in a dogfight over Saratov.[40] The Russians went on to have five female aces, and an entire regiment of women aviators: the 588th Night Bomber Regiment, made up of female pilots nicknamed "Night Witches" because they flew their missions in wooden biplanes at night. Many of the mechanics and bomb loaders of this regiment were also women. They flew over 24,000 sorties and dropped 23,000 tons of bombs. One woman flew combat missions when she was five months pregnant.[41]

Embittered Feelings

The way in which the WASPs disbanded caused much resentment and bitter feelings. WASP Barbara Poole wrote a scathing piece for *Flying* magazine's December 1944, edition entitled, "*Requiem for the WASP*":

We have spent large sums to train the WASP. Now we are throwing this money away at the demand of a few thousand male pilots who were employed,until recently, in a civilian capacity on government flight programs. The curtailment of the program has thrown these pilots out of work. And now they are to get the WASPs' jobs.

The demands of the unemployed male pilots have been a little far-fetched. "Throw the women out," they cried - meaning "and make room for us. They say, The women can't be drafted, but we can and will be.

But that's the very reason the women would have stayed where they were. What our Army needs most, our generals tell us, is men to hit on the ground. This is a sorry state of affairs for our pilots but after all we're running a war, not an employment bureau for disgruntled flyers.[42]

Many WASPs went home to face the rest of their lives on the ground. Despite heroic efforts, the WASPs found trust and acceptance hard to come by. After the resentment and bitter feelings subsided, they were left with the disappointment in not being absorbed into the Air Force. Many could not find a flying job. Helen Richey, depressed over the possibility of not being able to fly as a livelihood ended her life. Some former WASPs went to Alaska, or Mexico, and others went to Europe. Some went to jobs ferrying red-lined war surplus trainers, and some carved productive careers in civilian aviation. A few went home to laurels and tributes, but many others found anonymity, and marriages much changed and young men who did not understand why they turned away when war stories filled everyone's talk.[43]

None of the WASPs went on to jobs as pilots for the airlines or in the military, (although they all had the qualifications and the flight time). All of them received letters inviting them to take jobs as stewardesses. Some laughed at the letters, some cried, but most tore them up.

The WASPs volunteered to continue flying on a non-paying basis. After all, the war was still on and there was a need for skilled pilots. But there was no way the Civil Service would allow this.

Ten years after the war Jackie Cochran estimated that only 25 percent of the WASPs were still employed in aviation - many of these as flight instructors. Almost thirty years would pass until an airline hired a female pilot, and more than thirty years after they disbanded the WASPs would finally receive the long overdue recognition and veteran benefits. The benefits, however, were not retroactive. They did not get education benefits, insurance, or home loans. They should have gotten job preference like the men, but in fact did not. As far as medical benefits, only if the WASP was indigent did the V.A. step in. One WASP lost an arm to a B-26 propeller accident and the government declared her only 10 percent disabled.

Recognition

In 1976, the U.S. Air Force announced plans to train its "first women military pilots." To the WASPs the news was an outrage. They began a crusade to gain the long overdue recognition. They enlisted the support and power of Senator Barry Goldwater. Barry Goldwater was a major general in the Air Force Reserve, and during World War II had flown with the WASPs as a pilot with the Air Transport Command. He was the most experienced pilot in Congress with 12,000 hours in the air, in over 200 different aircraft. His voice would silence the opposition from the Senate. For the House battle the WASPs enlisted the aid of Colonel W. Bruce Arnold, the son of the late General Hap Arnold. Bruce Arnold, a strong supporter of Cochran and her WASPs, in turn enlisted the aid of Congresswoman Lindy Boggs, and Antonia Clayes, assistant secretary to the Air Force.

By 1977, the battle lines were drawn. Congress did not want to yield. Arnold and Boggs had one ray of hope in the House Veterans Affairs Committee to get their bill passed. The second ranking minority member was the committee's only woman, Margaret Heckler, of Massachusetts. A Republican, Heckler was co-chair with Democrat, Elizabeth Holtzman, of New York, of the newly-formed Congressional Women's Caucus. In March, through Heckler's efforts, the WASP bill became the only piece of legislation to be co-sponsored by every woman member of Congress

By May, the battle had begun. The Senate hearings took the full brunt of the "old guard." The American Legion testified: "In the history of our nation, the veteran has, from the time of the Revolution occupied a special place. It is highly prized and valuable, and is to be shared only by those who have earned it. To legislate such a grant of benefits would denigrate the term "veteran" so that it would never again have the value that it presently attaches to it."[44] The Legionnaires threatened to remain absent from discussions that advocated veteran's causes should the WASP bill pass.

The old animosities toward female pilots, and some say Jacqueline Cochran's ability to make long lasting enemies, died slowly. The Legion and other "man-made" roadblocks could not stop progress. On November 23, 1977, the eve of Thanksgiving, President Jimmy Carter signed the WASP bill and it became law of the land, more than 30 years after the WASPs had been disbanded. Cochran lived to see justice. The WASPs received the World War II Victory Medal and American Theater Medal in 1984.

They Made A Difference

In spite of this they had memories. Alyce Stephen Rohrer remembers: "Every time I flew at night I would hear music. I can't describe why. It was only the engine and the wind but it was gorgeous. It was a wonderful experience and I would not have missed it for the world."[45]

The WASPs made a difference to the war effort and for future women in military aviation. They would do it all over again if they had to. They blazed a trail for today's women in a time when there were no words like equal rights and women's liberation. They did what they did because it was the right thing to do. WASP Anne Dailey Marshall said, "These gals had a lot of guts - a lot of gumption - - - a lot of what it takes."[46] We must never forget them.

Chapter Four

The Daughters of Minerva

The concept of women in military combat roles is not new in the history of the world. The ancient Romans worshiped Minerva the goddess of wisdom, technical skill and invention. She is sometimes also referred to as the Goddess of War. The Greeks called her Athena.

The idea of America women in combat roles is not new either. There is documented evidence that American women served alongside men in every major military conflict both on American soil, and overseas.

Sybil Ludington

Women served in combat as ammunition bearers and riflemen, and supported the Continental Army during the Revolutionary War. Our history books are filled with the account of Paul Revere's ride. His ride the night before the Battle of Lexington in 1775 alerted Massachusetts colonists of the approaching British. It can probably be recited by most school-aged children in America. How many school children know about Sybil Ludington? On the night of April 26, 1777, the 16-year old patriot rode from town to town just like the older Paul

Revere, to announce to the Connecticut and New York colonists that the British had begun to raid Danbury, Connecticut. She was able to produce enough volunteers to repel the British the next day. Ms. Ludington's ride covered double the distance of Revere's, but somehow missed the school books.[1]

Deborah Sampson

Women also disguised themselves as men to get the job done. In 1782, a 22-year old woman from Massachusetts disguised herself as a 17-year-old boy. She distinguished herself for bravery as a ranger and a general's aide-de-camp in the Continental Army. Despite constant fear of discovery, Deborah Sampson, AKA Robert Shurtliff, was an excellent soldier and successfully adjusted to the war's primitive conditions, training and battles. Deborah served with the 4th Massachusetts Regiment in 1782, and was wounded in battle twice, first by a saber in a skirmish near Tarrytown, N. Y., and then by a musket shot near East Chester. The second wound was serious but she attended to it herself rather than risk revealing her true identity. The discovery of her ruse by a doctor treating her for a fever near the war's end brought praise from the military and the first congressional award of a full war pension to a woman, in September of 1818. Sampson later married and lectured about her experiences.

Molly Pitcher

During the Revolutionary War, the Battle of Monmouth, in New Jersey, served to provide a tremendous boost to the morale of a young nation. It also provided Americans with an enduring heroine, Molly Pitcher, whose legend still inspires patriots.

Molly Hays was a red-haired, freckle-faced Pennsylvania soldier's wife. It is not known for certain whether her husband was killed or wounded at her feet, but all accounts agree that she labored under the scorching sun to carry pitchers of water (hence her name) to the parched troops. When her husband fell wounded she took his place beside the cannon and continued the battle.

In the Civil War, many women served as nurses, saboteurs, scouts and couriers. They led troops into battle as color bearers, blew up bridges, cut telegraph wires, burned arsenals and warehouses, and helped prisoners and slaves escape.

Prudence Wright

Shortly after the "Shot Heard Around the World," The Battle of Lexington, word got around that a contingent of enemy troops and spies was heading toward Groton, Massachusetts. Prudence Wright assembled a group of women dressed in men's clothing and armed with muskets and pitchforks. They stood sentry duty at the bridge just outside town and captured a British agent with battle orders from the Crown. They swiftly turned both the spy and the papers over to the American revolutionaries.

Dr. Mary Walker

Dr. Mary Edwards Walker received the Medal of Honor from President Andrew Johnson for her outstanding medical work on the battlefields of the Civil War. She was later stripped of the honor in 1917 on the grounds that the existing records did not show "the specific act" for which the award was being made. This was ambiguous at best. Her achievements with the 52nd Ohio Infantry were well-documented. Some feel her unapologetic alternative lifestyle was a factor in swaying the lawmakers of the early 20th century. Walker's supporters continued lobbying for her until the medal was reinstated in 1977, 60 years after the death of the woman who earned it.[2]

Sara Edmonds

Sara Edmonds fought in the first battle of Bull Run and performed vital espionage work for the North. She was one of approximately 400 women who disguised themselves as male Union soldiers to assume combat duty during the Civil War.[3]

The invaluable service of female nurses during the Spanish American War led to the creation of the Army and Navy Corps. When World War I erupted, nurses braved the front lines over and over again, treating the sick and wounded. Other women enlisted in the Navy and Marine Corps or worked as Army switchboard operators.

In World War I, the Navy discovered that there was no gender-related law prohibiting women from the enlisted ranks and they recruited more than 11,000 women as "Yeoman" or clerical workers, in the Naval Reserve.

Louise Thaden devoted an entire chapter in her 1938 autobiography, *High, Wide and Frightened,* to a fictitious story about two women flying combat missions.[4] She wrote about women pilots recruited into the Army because of a severe shortage of trained, experienced male pilots. The two young protagonists draw dangerous assignments, and worry about any personal failure reflecting on all women pilots. In the end, one of the women is maimed but despite this personal tragedy, the message is that an individual's injury should not affect women's right to participate. In addition, an injury in the line of duty for a woman should be seen as a socially accepted hazard, as it is with men.[5]

Women took on new military roles during World War II due to the urgency of protecting our nation. About 400,000 women served in all branches of the military, performing a wide range of assignments.

One woman who was in the background throughout most of her military career was Rear Admiral Grace Hopper. She was known to her colleagues as the "first lady of software." The Admiral was not an aviator, but she was, however, a brilliant mathematician who joined the Naval Reserve in 1943. She helped design the first large scale digital computer, the Mark One, a breakthrough aid to trajectory calculations, the mathematics that helped us successfully launch vehicles into space. She coined the phrase "computer bug" to describe errors in computer software functions. When she retired from the Navy in 1986, at the age of 80. She was the oldest active duty military officer in the United States, she died in 1992 at the age of 85.

In 1948, the Women's Armed Services Integration Act gave women a permanent place in the armed forces. Since that time, in Korea, Vietnam, Panama, the Persian Gulf, and in peacetime, women have served with honor.

Between 500 and 600 Air Force women served in Southeast Asia during the Viet Nam War. In terms of numbers, the flight nurses were the largest group of military women in aviation to serve in Vietnam.[6] Operations officer Major Norma Archer gave the daily briefings on air strikes for the senior staff of the 7th Air Force. Air Force Intelligence was one of the most important functions performed by the women who served in Viet Nam.

When Da Nang Air Base came under attack, Lieutenant Jane A. Lombardi acted swiftly and deliberately in evacuating dozens of sick and wounded from the base. She was awarded the Air Force Bronze Star, and became the first combat-decorated woman in American history.[7]

Lt. Col. Yvonne "Pat" Pateman a former WASP, is another example. Her service record reads in part, "Bronze Star for exemplary service in Viet Nam; the Meritorious Service Medal for outstanding performance of duty with the Defense Intelligence Agency; the Joint Service Commendation medal for her exceptional performance with the Alaskan Command; the Outstanding Unit Award ribbon with one Oak Leaf cluster; and added the bronze "V" device for valor during her tour in Viet Nam; the Viet Nam Service medal with four battle stars; the Republic of Viet Nam Gallantry Cross with Palm Unit Citation, also for service in the combat zone.

Title 10 - United States Code
Although Pateman and other women had formidable records that showed they were exposed to combat, and performed heroically the government refused to allow women pilots in the military. There is no law that prohibits women from "serving in combat." The statutes which do not apply to the Army, specify only that women may not serve on ships and aircraft engaged in combat missions. However nothing in the law bars the services from applying combat exclusions to units other than ships and aircraft, and the services have done so. More important the law does not define "combat mission." The task is left to the discretion of the Department of Defense and the individual services, and how they define combat determines what jobs are closed to women.

In the Air Force Title 10 U.S.C. 8549 states, "Female members of the Air Force, except those designated under section 8067 of this title, or appointed with the view to designation under that section, may not be assigned to duty in aircraft engaged in combat missions." The exceptions designated under Section 8067 are medical and dental professionals, chaplains and other professionals.

Title 10 U.S.C. 6015 covers the Navy and Marine Corps. It states, "Women may not be assigned duty on vessels or in aircraft that are engaged in combat missions nor may they be assigned to other than temporary duty on vessels in the Navy except hospital ships, transports, and vessels of similar classification not expected to be assigned to combat missions."

In addition, the Marine Corps policy prohibits the assignment of women Marines to any unit within which they would likely become engaged in direct combat operations with the enemy.

No statute restricts the assignment of women in the Army. The Statute that covers the Army, Title 10 U.S.C. 3012, gives the Secretary of the Army authority to determine personnel policy for the Army. The Secretary of the Army has developed policies that exclude women from "routine engagement in direct combat." The Army justifies its exclusionary policies as being consistent with the implied Congressional intent (which is explicit in the Navy and Air Force exclusionary statutes).

This was a way of containing the women eager to serve their country as military aviators.

On March 25, 1966, Ensign Gale Ann Gordon, (USNR), became the first woman in the history of the Navy to solo in a Navy training plane, a T-34 Mentor. Ensign Gordon had an M. A. in experimental psychology and was a member of the flight surgeon's class at the Naval Aerospace Institute, Pensacola. Since she would be working with pilots, part of her training as an aviation experimental psychologist was to learn to fly. She did not become part of an operational unit.[8]

The issue of women pilots in the military continued to be debated, but debate was all that happened. Little action took place until 1973. Admiral Elmo Zumwalt, then Chief of Naval Operations, expected ratification of the Equal Rights Amendment, and he urged then Secretary of the Navy, John W. Warner, to open flight training to women. A voluntary program, he thought, would gain more support than a congressional mandate. Warner announced aviation training for women would begin, and their entrance into flight training marked the first time American women would be trained as military aviators. The initial group consisted of four women officers on active duty and four women officer candidates. Of this group six went on to become Naval aviators.

The Navy's objective was to determine the feasibility of using women in future non-combat flying roles, like helicopter and transport squadrons (Jacqueline Cochran's 30-year-old idea finally received serious attention.) After the women had completed their training, received their wings and served six months on flying status, an evaluation of the program would determine its level of success and its future. The program succeeded and since then, hundreds of women have successfully completed Navy flight school.

The Air Force took three years longer than either the Navy or the Army to incorporate women into its flight training programs. It maintained that even though women had completed programs in the other services, the reason for exclusion was not that the Air Force did not believe women could be pilots, but they claimed that the legal restrictions on women in combat made it difficult for them to justify an extensive program to bring women into Air Force flight training. The Air Force claimed that all its pilots were (potentially) combat pilots, although many pilots spent their entire careers as instructors, transport pilots and other assignments without drawing a combat tour.

Before the first half of the first class of women had even finished training, the Air Force decided to continue its program. Another nine women began training in February 1978. The reason for this was the positive reactions of the instructors. Lieutenant Colonel C.T. Davis, operations officer for the 96th Flying Training Squadron reported that: *A lot of people had a lot of ideas that women wouldn't be able to hack it because of their lack of physical strength, because of inadaptability to stress, because of this, because of that. So far our experience has been that it hasn't changed things at all. The women are going through exactly the same training as the men and are hacking it just as well.* [9]

Operation Desert Storm

In 1990, women's role in the military took on a new dimension. They lived like pioneers - toughing out sandstorms, being chilled to the bone at night, and roasting during the day in the hot Saudi Arabian desert. Others served their time in more comfortable conditions on aircraft carriers, but all of the women pilots who flew in the war have their own personal stories to tell; some happy, some sad, but all memories to cherish.

The women functioned at peak performance in a culture that routinely places women in secondary or subservient roles. The U.S military complied with local Moslem customs and restricted women's movements off the base. They were not permitted such routine basic necessities as driving an automobile.

They washed their hair with bottled water, endured two-hour dusty truck rides just to make a phone call home, and waited like any other soldier for mail from home. They were Regular Army, Air Force and Navy and they were reservists, one day a civilian the next day doing what they were trained to do - fly. Now women can, if they choose, tell war stories once the province of men only.

Lt. Cmdr. Trish Beckman - NFO F/A - 18

Lt. Cmdr. Trish Beckman has seen many changes in her 22 year career. Beckman had only a high school diploma when she entered the enlisted ranks as an 18-year-old. As an enlisted person, Beckman was a flight simulator operator and as part of her job she performed flight attendant duties on VIP flights. The Navy let the women fly the same hours and missions as the men, but at non-crewmember pay and no wings. That situation had many women angry, including Beckman, but Beckman did something about it. "One day I asked, why can't I be a permanent crewmember and get my wings?" The brass replied, "Women can't do that." Her next question was, "Where is that written down, sir?" After several weeks of research, the brass came back to her with an answer. They could not find it in writing. Not long afterward, Beckman and all the others got their wings, permanent crew status and crewmember pay. "Imagine that," said

Beckman smiling. "I learned to keep asking questions until I got the answers I wanted to hear." See Fig. 4-1.

Beckman went on to college, officer candidate school and then on to Pensacola for flight school. She is one of the first female Naval Flight Officers to graduate. After earning her wings in 1981, she went on to a tour of duty in Hawaii. Then it was on to graduate school and a masters degree in aeronautical engineering. From there she earned a slot in the Navy's prestigious test pilot's school.

Lt. Cmdr.Trish Beckman is currently the vice president of the Women Military Aviators, and has been at the forefront of women's rights in the military. The 1991 Tailhook sexual abuse incident against 26 women, over half of whom were Naval aviators, inflamed the feelings of many women in the military, and Beckman spoke out about the episode.

"The sexual assault at Tailhook was a direct result of a small number of men feeling threatened by the repeal of aviation combat exclusion laws. Their fear was that they would have to compete with women on a level playing field. It was an attempt to put women aviators 'back in their place,' in the same way that rape and physical violence have been used by men throughout history to keep all women 'in their place.'

"What happened at the Tailhook convention was a crime, and aiding and abetting is also a crime. There are some in the Navy leadership who continue to condone such criminal behavior and say, It's being done just in fun to women who asked for it, otherwise they should not have been there.'

"Tailhook has become the catalyst for social change in the Navy. It will be the stimulus for the equal treatment of all military women, because the other services are watching. Coupled with other recent major events in this country (Clarence Thomas Supreme Court nomination, William Kennedy Smith, and Mike Tyson rape trials), it will be a major factor in the political climate for years to come."

As a direct result of Tailhook, all Navy personnel received mandatory sexual harassment training in 1992. Until then, there was a belief among the troops that the zero-tolerance policy for substantiated sexual harassment would not be enforced, since their superior officers participated in it, and their commanding officers condoned such behavior. "The zero-tolerance message is now coming down loud and clear from the top brass," said Beckman, "and with proper emphasis, the message will get through to everyone, but only if the admirals enforce it and provide the appropriate example.

On September 24, 1992, Acting Navy Secretary Sean O'Keefe swore a new commitment to solving a "cultural problem" of demeaning behavior toward women.

"We get it," said O'Keefe, who added, "those who don't get the message will be driven from our ranks."

With those words he announced the forced retirement of two Navy Rear Admirals, and the reassignment of a third. Retired were Rear Admiral Duval

Fig. 4-1 Lt. Cmdr. Trish Beckman

Williams, Commander of the Naval Intelligence Service and Rear Admiral John Gordon, Judge Advocate General.

"The cultural problem of gender bias has to be methodically eliminated," said O'Keefe, describing the Tailhook incident as "behavior unbecoming human beings."

"Because of Tailhook," says Beckman, "some people think that the Navy has a more macho attitude than the other services. This is not true. The Navy stands ready to implement women in combat aircraft, whereas the Air Force and the Department of Defense are holding back."

Lt. Cmdr. Beckman sees military women on the verge of total integration into combat aviation. "We have the experience of Desert Storm and the repeal of the aviation combat exclusion laws in 1991. Against us, we have an administration which intends to delay implementation as long as possible. Those who are against us are mostly well-meaning, but when they decide what is best for women as a class, they treat me as second-class citizen.

"It is well past time for a change in the treatment of women in the services," she said. "Historically, any major social change is preceded by violence. This may be the first time in which we affect major social change without violence. I do not expect or advocate violence, but military women should not tolerate such treatment. I have great respect and admiration for Lt. Paula Coughlin, who spoke out against the sexual assault she encountered at Tailhook. She deserves the credit for bringing about a positive social change in the military."

The other side of the sexual harassment coin is the pseudo-protector. "Some well-meaning men believe they are protecting women by not allowing us to compete for 'combat'jobs," says Beckman. "If these men really believe that they are the "natural" protectors of women, then why do American men commit such a wide range of violence against women daily in their own homes? And why do servicewomen encounter more violence from their own countrymen in peace than while they are prisoners of war of other countries? Do these men really believe that it is OK for American men to rape American women, but is not OK for men of other countries to rape American women?

"What are they protecting us from--equality of horror? Rape, mutilation, and murder in our own homes can't be that much less horrible than in a POW situation. At least POWs are protected by the Geneva Convention and world opinion. In modern warfare, the worst horrors of war are more often witnessed by military medical personnel (many of whom are women) and suffered by non-combatant civilians (most of whom happen to be women and children).

"Military women are tired of someone else determining what is best for them," says Beckman. "I am qualified and capable of performing equally with men in combat aviation. Full integration of military women into combat aviation will quickly dispel any myths about our capabilities, and will reduce sexual harassment to a minimum.

"The excuses for excluding women from combat are the same 1940's excuses to exclude blacks from full participation in the military, e.g. 'they will destroy unit cohesion.' The exclusionary tactics were sanctioned as a means to control blacks and to 'keep them in their place.'" (In the early 1970's, when blacks rioted and set fire to a few ships, the Navy leadership finally got the message that such treatment was intolerable. The Navy initiated behavior modification programs, which educated people at first and now punishes those guilty of racial bigotry.)

"The same methods are apparent today in the sexual bigotry in the Navy. Women are limited to certain duties, physically separated from their teams, and subjected to verbal and physical assaults.

"Any social change brings an adjustment period. Military women should brace themselves against the same resentment that blacks endured. Patriotism, dedication to duty, and perseverance are qualities that military women obviously possess because they remain in the service in spite of the widespread sexual harassment and discrimination. The fact that women already possess those qualities will serve us well in the coming adjustment period, and make us stronger. We must not be afraid to demand what we are qualified for."

Lt. Cmdr. Beckman believes the majority of Navy men believe in the American ideals of equality. However, a few are quite skillful at providing a hostile atmosphere for Navy women. "The Navy leadership has done little to punish those sexual bigots, in the past," says Beckman.

The key to the acceptance of women in the military is the repeal of all combat exclusion laws and the associated service policies. Two-thirds of the American people approve of women being allowed to volunteer and compete for jobs they are capable of performing (i.e., law enforcement, firefighting, drug enforcement, and combat aviation).

"I believe we will see a woman as Chief of Naval Operations within 20-30 years," says Beckman. "I see a brilliant and (mostly) equal future for military women in general and for military women aviators in particular. The Navy will be the first in this area. It led the way for women in the past. Navy women officers were air navigation instructors and flew as technical observers in World War II. They were in pilot training in 1973, flew A-7's in 1975, and were on ships in the late 1970's. They taught men to dogfight and bomb targets in the late 1970's, and were the first in test pilot training in the early 1980's."

Cmdr. Beckman believes restrictions must be removed, and women must fly in combat roles or drive combat ships before her prediction will come true. "For the young woman thinking of a career in military aviation--begin early to prepare yourself mentally and physically for the challenge. Take the most technical curriculum available in school, get involved in team sports, and work out with weights. Don't be afraid to be aggressive and stubborn, because those qualities are admired in military aviators. Look forward to the day when you will compete equally with men for the ultimate war games in military aircraft. And remember that in the military you may be required to pay the ultimate price for your country."

Capt. Ann Patrie - UH-1 Pilot

Captain Ann Patrie has been flying since she was 13, and she could not tell you what life would be like without an ID card. Ann, the daughter of an Army communications and intelligence officer traveled throughout the world as a child. She settled in Texas and spent her teen years in San Antonio. Her grandfather was one of the first members of the CAA and he was in charge of airfields and airways. Her father was named after Charles Lindbergh, "Lucky Lindy," so Ann's background has always had an affiliation with aviation.

Ann was a military dependent until she graduated from college and joined the Air Force through an ROTC scholarship. Her first assignment, as a second lieutenant, was at Keesler Air Force Base as an air traffic control officer. From there she went to Myrtle Beach, South Carolina for upgrade training. It was there she found out that there were too many air traffic control officers and the Air Force made her a logistics officer. After moving through several bases, Ann decided she wanted something more than pushing paper. "I felt the Air Force had not tapped into my real potential.

"I have always been adventuresome but unfortunately the Air Force is very technical and they put you in a pigeon hole and you really don't have a lot of latitude. I wanted to go into a different career field but the Air Force would not

Fig. 4-2 Capt. Ann Patrie

transfer me. I had spoken to the Army about their aviation program and their age requirement was a maximum 30 years old. The Air Force told me I was too old for flight school at 27."

Ann left the Air Force and got a commission in the Army Medical Service Corps in March 1987. She graduated with her wings in April 1988. Capt. Patrie flies a UH-1 medivac helicopter, the one with a big red cross - which makes a great target. See says.

By November 1990, Patrie had joined a reserve unit and was off on a job interview in San Antonio when her commander called. Her whole unit, 15 pilots, a full compliment of crew, crew chiefs and flight medics had been activated for the upcoming Desert Storm conflict. She was to report back immediately.

Capt. Patrie was the only woman pilot in her unit, the 374 Medical Detachment and in the entire 807th Medical Brigade. There was one female flight medic and six other enlisted females in crew dispatch and other support roles.

Her unit had eleven days to process, train and do whatever else was necessary to meet combat-ready requirements.

Thousands of troops were "shipping out" and overnight stops at air bases usually resulted in overcrowded accommodations. Sleeping bags and hardwood gym floors were the norm. Patrie's knowledge of the inner workings of the Air

Force was useful in appropriating comfortable quarters and creature comforts for her unit at the two stops they made in the United States. See Fig. 4-2.

"Initially we were told we were going to Bahrain, but other units had got there first, and all the good assignments were gone. The units coming in would be deployed farther and farther forward and the closest units would probably support the initial combat engagements.

"We were really surprised that here we are, a little podunk unit from Little Rock, Arkansas, and we were one of the most forward deployed reserve units during the war. We wound up going to support the 3rd Armored Cavalry Regiment, which would be the second wave to face off with the Republican Guard during the war. We were located eight miles from the Iraqi border.

"We were warned Saddam was going to gas our people and the medical people were going to be prime targets. Never in my wildest dreams did I ever think I would go to war. Me? A reservist, a woman, go to war, it just doesn't happen."

Capt. Patrie described the living conditions as austere. "We lived in a tent from the day we got there until the day we left. I have never been so dirty in my entire life." They did not have the luxury of separate facilities for women either. "I lived in a command module tent with my commander, my operations officer, the other section leader and the first sergeant. It was very crowded but the gentlemen respected my privacy and I respected theirs. We were able to cordon off a space for me. At night when we slept I closed my area off, it afforded me a little bit of privacy, my own little space. If I changed, or they did, we closed the curtain. We were all very professional."

At first Patrie's unit did not have enough heaters, so as Capt. Patrie puts it, "We had to go on recon missions to acquire more. There were men waking up at two in the morning because they were shivering."

The reaction from the Saudis toward women was predictable. "We had very little contact with the local people. The Saudi's would not do business with the American women. Contacts were mostly with our own troops and support units."

One mission Patrie and her crew had was to pick up human blood. "We flew through enemy territory, we flew very close to the desert floor to avoid drawing enemy fire. After picking up the blood, I flew over Jabail Air Field in Iraq. I flew up and down the highway to look at the carnage left behind. The road was littered with trucks and bodies. I saw where the smart bombs had gone through hangars and blown up aircraft sitting in them. I never saw such destruction. I don't know if any other female went that far in supporting a combat role." Capt. Ann Patrie personally logged more than 30 hours flying combat.

"For the first time in my life I saw fear manifest itself in different ways in the men."

Capt. Patrie spent five and a half months in the Saudi Desert as a Medivac pilot. Her crew just happened to be equally divided along gender lines. Capt. Patrie and Staff Sgt. Laurie Cone made up one half of the crew with a male co-pilot and crew chief filling in the other seats. During the war, Capt. Patrie and her crew penetrated more than 100 miles into enemy territory.

Patrie is married to Capt. Christopher Patrie, a C-130 pilot instructor who was not sent to Saudi. He found that incredible but accepted the decision and strongly supported his wife. When she came home, she was a hero to her husband and the rest of America.

Staff Sgt. Laurie Cone - Flight Medic UH-1

"I joined the Army Reserve at the age of 27," said Staff Sgt. Laurie Cone, "because I was so impressed with the price of freedom and the privilege to be living in America. I thought I could give of my time and become trained as a citizen-soldier so that if the need ever came, I'd be prepared to help instead of panic or need to be looked after." Laurie has been a flight medic for the past seven years. At any moment there are numerous opportunities for a qualified flight medic to take special assignments nearly anywhere in the world. Laurie has had some extended tours in Central America. See Fig. 4-3.

Laurie chose the combat medic field because she saw that as useful in everyday life as well.

Air ambulance was looked at by ground medics as the highest level of life support in the field environment. Laurie felt that the basic level of medic training did not equip her to perform as a competent flight medic. To overcome this difficulty, Sgt. Cone sought civilian training in the emergency medical field and took every course the Army offered. Today she is skilled in many advanced life support techniques and classifications.

As the only female flight crew member in her unit, Laurie faced some prejudices from her male counterparts. This difficulty was enhanced by the fact that flight medics and crew chiefs on the UH-1 Huey and the UH-60 Blackhawk are cross-trained so that they can assist each other with crew responsibilities.

"The crew chiefs thought the job was for men only and consistently tried to put pressure on me to fail," she said. "I was so eager to learn about aviation that I enthusiastically took every assignment they gave me. I volunteered for jobs no one wanted to do - I'm sure that the many trips to the hangar for tools and fuel sample bottles were invented to test me as well."

There was always speculation about the role of Laurie's unit in a combat environment, especially since it had a female. "Of course you'll be replaced by a male medic and have to stay at a ground hospital somewhere in the rear," said the men. "I never openly argued with the guys," said Laurie, "but I was secretly determined to be of such value as a crew member that they had to include me.

"I took every mission I was offered, however minor, so that I could build up my crew skills. This included night vision goggles training, and water survival for flight personnel." When Laurie's unit was called up for Desert Storm, she was the most qualified medic in the unit.

Sgt. Cone fell in love with aviation through her contact in medivacs. "The versatility (of rotary aircraft) is incredible," she said. "I have unofficially flown

Fig. 4-3 S/Sgt. Laurie Cone

several hours and I'm nearly obsessed with flying. I plan to get my pilot license after graduation from college in 1993."

The aviation field has provided me with the opportunities to apply my energy and enthusiasm. Once the guys saw that I was on the job to be professional, they became very supportive. Many of them have even requested that I be on their crew because they could count on professionalism."

Sgt. Laurie Cone also had the opportunity to compile many ideas for training flight medics into the Crew Readiness Training Guide that was adopted by the 5th Army to be used in medivac units as a standard.

"I believe that because I maintained a don't get your honey where you get your money' rule, my work atmosphere kept a lot of pettiness to a minimum. I believe the men I served with appreciated this even though they tested me often."

During Desert Storm Laurie lived in a tent 12 miles from the border of Iraq with 11 men, and she deployed over 100 miles into enemy territory.

"I can say without a single reservation women have an important place in combat. I don't believe women should be reduced to just a support role. There are many roles they are fit for - aviation is a good example.

"It takes a special kind of woman to serve in the military. She must be there for the right reasons and have the determination to work hard in spite of the stereotypes put upon her. She must have the heart for serving her country. This asset will carry her when all the odds are overwhelming."

Sgt. Laurie Cone would also like to see the United States adopt the attitude Israel has regarding its women. Their top sharpshooters and pilots are women. "And their uniforms are not made to disguise their beauty."

What's in Laurie Cone's future? "Serving my country in wartime is the greatest privilege to date. The next great adventure of even greater magnitude for me will be having the opportunity to shape the future as a school teacher."

Lt. Kelly Franke - 1991 Navy Helicopter Pilot of the Year

A 1983 graduate of E. J. Wilson High School, Lt. Franke was pursuing a major in psychology at the University of Rochester, with intentions of going to medical school. "I started as a pre-med student, but that was unsatisfying. I began to dream of flying again. As a small child in Spencerport, New York, I had always dreamed of flying." After researching different routes, Franke enrolled in the university's Navy ROTC program, and graduated with a bachelor's degree in 1987. She joined the Navy in May of that year and attended flight school in Pensacola, Florida. She received her gold wings at Naval Air Station Whiting Field in April 1989.

"I was told when applying to flight school that there weren't many billets for women, and I should apply to the Supply Corps instead!" Franke ignored the advice and her first assignment was with Helicopter Antisubmarine Warfare Squadron One in Jacksonville, Florida.

In 1990, Lt. Franke deployed with the "Desert Ducks" to Bahrain. She spent a total of twelve months there, on two separate deployments in Operation Desert Shield and Desert Storm. See Fig. 4-4.

"When the war began, questions arose over whether the female pilots should be allowed to stay in Bahrain. We convinced them that there weren't any good reasons for us to leave."

During Operation Desert Storm, U.S. navy Carrier battle groups launched round-the-clock air strikes against Iraq.

Navy pilots and aircrews received hundreds of awards for heroism and daring in direct combat action. But, no less daring were the Navy helicopter pilots and aircrews, who provided the vital logistics support necessary to keep those battle groups operating. It wasn't a glamorous job. Keeping the battle groups supplied day or night, under any and all weather conditions, was at times harrowing.

Franke's missions often involved flying in hazardous conditions in the Persian and Arabian Gulf. During Desert Storm Lt. Franke flew numerous logistics runs dangerously close to Iraqi-held oil platform gun positions. On her second deployment in June 1991, she found that flying conditions were worsened by environmental hazards, with the intense summer heat, and skies darkened by countless oil fires.

Fig.4-4 Lt. Kelly Franke

She described the typical day as a lot of hard work, with missions averaging 8-10 hours each. Yet, she enjoyed the deployment. She has nothing but praise for the "camaraderie and the professionalism" that makes her squadron special.

Lt. Franke earned both regional and national recognition as 1991 Pilot of the Year by the Naval Helicopter Association (NHA). Cited for "inspirational" and "unparalleled" performance, Lt. Franke flew 664.2 accident-free flight hours in 105 combat support missions. These missions included transporting Iraqi prisoners of war, and numerous logistics runs to the battle groups before Iraqi gun positions had even been cleared from sea-based oil platforms.

On one occasion, she led her crew on a night mission to rescue a Navy diver from a mobile dive platform in the Arabian Gulf. It was impossible to land on the floating platform, and she hovered over the platform at 75 feet in a 20-knot tail wind, while the crew hoisted the injured diver to safety. See Fig. 4-4.

Three months later, she averted disaster during an in-flight emergency in the northern Arabian Gulf. Supplying several ships that day, her crew had off-loaded supplies aboard the Navy frigate USS Elliot, their third delivery for the day. Ten minutes after leaving the ship, their helicopter lost auxiliary hydraulic pressure.

Loss of auxiliary pressure requires the pilot to land as soon as possible, and with the nearest land 40 miles away, Lt. Franke decided to make an emergency landing on the frigate USS Stark. Battling for control of the aircraft in a 30-knot crosswind, which robbed the helicopter's tail rotor authority and forced the nose of the aircraft down, Lt. Franke aborted her attempt to land on the small deck. She skillfully recovered from a potentially disastrous spin to the right and made a safe return flight to Kuwait, escorted by another ship's helicopter.

The escort pilot witnessing the incident praised Franke's "immediate and decisive" handling of the situation, allowing her to regain control of the aircraft in seconds, saving the aircraft and crew.

Lt. Franke does not fit the common perception of a tough, combat-seasoned pilot. Squadron officials say 26-year-old Kelly Franke has a well-deserved reputation for a prowess and savvy in the behind-the-scenes role of keeping tons of supplies on the move to combat forces. The soft-spoken, self-assured pilot won the praise and admiration of officials, who felt she was "a must" for the prestigious title of Pilot of the Year.

She described her Desert Storm experience as "tough" and "demanding," but one she was proud to have experienced. In fact, the Navy has given her an opportunity to do the one thing she has always wanted since she was a little girl in her small western New York hometown.

Lt. Franke wears the Navy Achievement Medal, Southwest Asia Service Medal, a Sea Service Deployment Ribbon and the Kuwait Defense Medal. She admits that she was surprised, however, to win the Navy Helicopter Pilot of the Year Award. "I was doing my part (over there)...just like everyone," she says. "If I had to do it again, I would."

Lt. Franke has never regretted her decision to join the Navy, and plans to make it a career. Like the adventurer she is, Lt. Franke is anticipating greater challenges in her next assignment...perhaps as a helicopter pilot in Antarctica, or the Navy's test pilot school.

"The military, with all of the changes it is experiencing, is where the greatest opportunities for women exist," said Franke. Women really can go as far as they want in the military, and I don't believe it will be long before it is totally equal for men and women. It is no longer an issue of I'm a women, and I'm out to prove something.' We don't need to do that anymore. Through my short career, I've felt that if you want something bad enough, and perform, and set the standard, you are going to attain that goal.

"The goals of equality for men and women are within reach. We, as women, simply need to show that we are there to accept the important positions with our male peers."

First Lt. Ellen Ausman - Navigator EC-130

Air Force First Lt. Ellen Ausman talks seriously of her flight into Kuwait City. The situation she describes was serious. "We had a cargo on board that was really important. The smoke from the oil field fires was so thick that the control tower at Kuwait Airport was closed because they couldn't see the runways." Landing at this field was at a pilot's discretion. "We made the decision as a crew

Fig. 4-5 Lt. Ellen Ausman

that we were going into a field that was not lighted, there was no tower and their traffic was IFR under VFR conditions." She explains, "You're flying on Instrument Flight Rules but the traffic is managed under Visual Flight Rules, and you can't see a thing. It's a very tricky situation."

The crew decided to set up a navigator's approach but Lt. Ausman was a visitor on the flight so she volunteered to help with the chart reading. See Fig. 4-5.

The navigator picked a ground target on his radar scope, and plotted a course to fly the plane around any obstacles and down onto the runway. "At five hundred feet we should have seen the runway," said Ausman. "We had five sets of eyes looking out the windows but we couldn't see it. The smoke was so thick that all you could see was straight down.

"All eyes were straining when a voice called, "You passed it. Let's go around.'"

The pilot pushed in the power and they flew back out to a long final to try again. By then the crew had turned on the cockpit lights. The smoke was so thick that it was like nighttime. The copilot was flying the airplane and the pilot was looking ahead for the runway. Suddenly the pilot spotted a building with a crane on top of it. The co-pilot added power and pulled up and over the building. He then dropped back down to his approach.

Lack of visibility had caused some serious confusion. "About two-thirds of the way down the runway, we think we're lined up with the right runway, but we're really lined up with the left. The pilot yanks and banks and bumps it in."

As it turns out, they could not land on the right runway. The first 3,000 feet were closed because of unexploded bombs stuck in the concrete.

"Once we were on the ground the atmosphere was so dark with soot that you could look straight at the sun and it was like looking at a full moon on a dark night."

The crew was cautioned before they deplaned that the area was littered with anti-personnel devices. The only clear area was the left runway, the ramp and the hotel area. So much for "sightseeing."

Lt. Ausman's job as a navigator is to position the aircraft precisely in the sky, using the onboard navigating equipment, the stars and computers. Throughout the war, there were EC-130s flying non-stop from January 17 until the cease-fire, providing aerial reconnaissance, electronic jamming and air traffic control.

When Lt. Ausman and her crew arrived in the Arab Emirates in September 1990, they found instead of barracks, a tent city. Although the tents were air-conditioned, they were not airtight and sand was a constant problem. Sleeping space was also at a premium. Ellen had to "room" with 28 other women.

"The latrines were rows of toilets with only about a two-foot-from-the-wall curtain between each one. No doors and not much privacy - but better than an outhouse."

Ellen Ausman first became aware of restrictions on women in the military when she joined the Air Force ROTC at Colorado State University, and since then she has had her share of career difficulties. In November, 1988, she began specialized undergraduate training at Mather Air Force Base. She was the only female in the 80-student class. "In August 1989, I graduated in Class 89-13, the only female in 60 graduates," she said smiling.

Ausman credits the women who went before her for blazing the trail and opening some of the doors. Their impact on the "system" has made her enlistment a little easier and she hopes her presence will make a difference for the women who follow her. "No single woman can do it all. It takes all of us. Change is incremental, and together we can affect change."

Lt. Ausman and the other women in Desert Storm also learned something about the Arab culture. For example, she couldn't just get in a car and drive around the countryside. Women don't do that in Arab countries.

"The Arab women would glance at us, but the Arab men would stare at the American women non-stop. If an American man stared at an Arab woman he could be arrested. All she had to do was feel insulted, raise her hand and the police would arrest the man. It was that quick. Yet their men could stare at us and we couldn't do anything about it.

"I enjoyed being surrounded by our men," said Ausman. "We couldn't fraternize but we did have fun," she says as she flashes a wide grin.

But there was never any quiet. There were always planes starting their engines, or shutting down or zooming overhead. And there was constant truck traffic.

The weather was also hostile. On summer days the temperature went up to 120 degrees and in the evening it went down to 100. In the winter it was around 75 in the daytime and 35 at night.

Lt. Ausman flew 32.5 combat hours and during that time she had an actual high-flying birds-eye view of the raging battle. "I saw enemy anti-aircraft fire coming up at us, and the B-52 bombers that came right over our heads and dropped their bombs on the targets. I saw F-15s, F-16s, F-18s and A-10s all doing their missions. We could see missiles firing right beneath us and the bombs going off below. We could see the fires and smoke from the oil wells in Kuwait.

"The biggest threat to us was our own friendly aircraft. At any time when I looked out the window I saw eight or nine planes flying around. We didn't move to get out of the way until we saw which way they were going to turn. There were aircraft up there all the time. It was non-stop adrenalin."

Ellen is back home now and in her spare time she flies her 1946 Taylorcraft. She proudly displays it at airshows. Ellen got her private license in 1987, and her instrument ticket in 1988. She credits her cousin and mentor, Jean West, with teaching her to fly.

Lt. Ausman is a motivated person who sets goals for herself. One day she hopes to pilot the EC-130 that she now navigates.

To any young woman thinking of a career in the Air Force, Lt. Ausman says, "Go for it - you'll make me proud."

AE1 Sue Patelmo - USN Search & Rescue Swimmer/CH-46 Crew Chief

The Navy offers a wide range of career opportunities for women, many of them in the enlisted ranks.

Sue Patelmo enlisted in December 1979, and her first station was NAS Cubi PT Philippines as an H-46 helicopter electrician. When she started flying as an air crewmember in 1982, there was no requirement to attend Naval Aircrew candidate school in Pensacola. In the middle of her flying that changed. So Sue left the Philippines to go to Naval Aircrew School. "I also qualified for Rescue Swimmer School but was told I could not attend because it was closed to women and considered a combat billet. I really wanted to be a rescue swimmer. From Pensacola I received orders for Antarctica, as a C-130 loadmaster.

"From there I became an H-46 aircrew instructor. During that time (1985-1989) Rescue Swimmer School was open to females. I went to Rescue Swimmer School and became the first female rescue swimmer in the Navy. I believe there is another female rescue swimmer. She is stationed on the East Coast. We are the only two in the Navy." See Fig. 4-6.

Sue Patelmo deployed on the USS Shasta during Operation Desert Shield and Desert Storm, working as crewchief, Natops Instructor, electrical and avionic trouble shooter. Sue is also a quality assurance representative who

Fig. 4-6 AE1 Sue Patelmo

inspects maintenance tasks and removal and installation of integral systems of the H-46 helicopter.

Right now Sue Patelmo is a Squadron Natops Instructor and responsible for assuring all aircrewman maintain their qualifications and administering tests and Natops checkrides. She evaluates new crewchiefs and requalifies current crewchiefs.

Sue had some early career difficulties, like waiting years to go to Rescue Swimmer Schools. Consequently she missed out on a re-enlistment bonus and the chance to do the thing she wanted most. Today things are changing.

"There are many more opportunities for women in the Navy. Much more so than twelve years ago when I first enlisted. If women are diligent in their pursuits and set goals for themselves, they can have a noteworthy career and a sense of personal satisfaction knowing that they have consented to serving their country, with pride.

"No matter where you are, in what profession, there are always going to be obstacles to women. In my twelve and a-half years of service in the United States Navy I have really worked hard accomplishing the goals that I have set for myself. I still haven't stopped.

"Being a part of Desert Shield and Desert Storm is something that I will never forget. The wide realm of emotions that you go through is incredible. Fear, confidence, loneliness, apprehension, etc. It's all a big part of war. Getting the job done when it seemed your body could just not move anymore. Living in conditions, well, you had to be there."

Lt. Cmdr. Catherine Osman USN CH-46 Pilot

Cathy Osman has been in aviation for the last ten years. "I applied for flight training while in NROTC. The board only met once a year then and quotas were limited. My records were incomplete, so I had to wait until June 1980, the following year to try again. I received my wings on August 14, 1981, and reported to HC-16 during November 1981. My first squadron had never had female pilots. I had no 'wing mate' so I wanted to avoid going anywhere that I had to be paired to deploy. The squadron in Pensacola worked off the Lexington HC-16, Search and Rescue support for the Lexington, the Navy's training carrier which had women as part of its crew. I thought that it would be easier to cruise there than someplace else. Never the base, I wasn't welcomed to the squadron with open arms."

Osman deployed during November 1990 through January 1991 in support of Desert Shield, and again in May - November 1991 in support of Desert Storm.

"The outlook for women depends upon the future of the military in general. Cutbacks will probably affect women more so than men.

"When I came into the military, I was in awe of the women ahead of me. What they endured so that the rest of us could follow in their footsteps was not just discrimination, but outright harassment, yet they persevered. The women coming in now are aware that they may have to work a little harder, but they expect to have the same opportunities as their male counterparts. At least, that is true in helicopter support. We are still restricted from the 'best' flying, not only in helicopters, but in jets. Regardless, we are an integral part of the HC community and should survive the cutbacks. I hope that women will continue to be treated as the norm in this community and that the other communities will soon follow suit."

Capt. Deborah Myers U.S. Army - UH-60 Pilot

Capt. Deborah Myers was commissioned in May 1985, and applied for aviation as her first choice and intelligence as her second choice. As the first female from Texas Tech in the ROTC program she was told she would never be selected. She was surprised and very excited when she was told to report to Fort Rucker for aviation training.

"The only difficulties I had were I did not think I was very mechanically inclined. I had never even seen a helicopter up close." Debbie did as well as her peers but, "There were some areas that I had to spend more time studying for than they did," she said. She was also emphatic about one thing. "I feel like one of the luckiest people in the world. The Army has treated me well. I have been given every assignment and aircraft I've asked for." See Fig. 4-7.

Fig. 4-7 Capt. Debbie Myers

Some of the assignments she has had include UH-1 qualification, and UH-60 qualification and transition, Military Officer transition, advanced courses, and fixed wing training.

When Debbie's unit was deployed to Saudi Arabia there were 135 soldiers including 12 women. "Every single person in my unit was a professional. Everyone pulled their share and did their job. The guys were great with only a few glitches. We were 'in the dirt' with the 1st Air Calvary for the entire time.

"The night of the air war I was working in flight operations and heard the report over the radio of the air raids. That is the proudest that I have ever been. I thought the Air Force pilots were great. I think they made it easier for the soldiers on the ground."

Capt. Myers remembers very clearly the feelings she had on her missions. "It was me and another female pilot and a male crew chief. As we were flying across the border it struck me that we would be in trouble if something went wrong with the aircraft. Especially with the Middle East views on women. (Some women military truck drivers were arrested in the Desert Shield deployment, for operating a machine.) Then too, the enemy was known to be vicious with prisoners.

"You learn a lot about sociology and psychology when you are in an environment like that. I found that you should never feel sorry for yourself because the person next to you always has it worse. You go through a whole range of emotions, fear, hope, happiness, sadness, frustration, anger, etc. I'm from a strong family and had prepared myself mentally before I left. If I needed to die for my country I was ready. I was ready, but I'm not a war-monger.

"I was shocked and really happy by the support from the people back home. I think that we owe a lot to the Viet Nam veterans. They paved the way, and I just wish it would have been different for them when they returned home."

Capt. Myers also had some advice for women today. "Never let anyone tell you that you can't do something. If you truly want it, set your sights on it and go for it. But also realize it may not be an easy goal. Just give it your best shot and take one step at a time. It will happen if it were meant to be, but you must give it 200 percent of your effort. Hard work and professionalism will always keep you up with your peers, if not ahead of them."

Cmdr. Lucy Young - C-9 Pilot

Lucy Young won a four-year Navy ROTC scholarship to Purdue University and graduated with a B.S. degree in 1976. After graduation she received her commission as an ensign and reported to Naval Aviation Schools Command, in Pensacola, Florida, in October 1976. A year later she had earned her Naval aviator wings and then went to jet transition at the Naval Air Station (NAS) in Kingsville, Texas.

After qualifying in the TA-4J Skyhawk, Young reported to NAS Barbers Point, Hawaii, where she accumulated over 1,000 hours in fleet support missions and multi-national exercises. Young qualified as a Section Leader, Instructor Pilot and Air Combat Maneuvering Pilot, and then received orders to Training Squadron, NAS Kingsville, Texas. As a flight instructor, she taught student Naval aviators in all phases of advanced strike training. Young was also one of the first women in the Navy to carrier-qualify, in May, 1982.

After leaving active duty in July 1983, Young accepted a commission in the Naval Reserve. In December, 1983, she accepted a position in Atlanta, Georgia, as the first female FAA Flight Test Pilot. There she performed flight tests on aircraft, avionics and navigation equipment for FAA certification. See Fig. 4-8.

In June 1986, she began training to become a Boeing 727 flight engineer with Piedmont Airlines. In October 1987, she upgraded to first officer on a Boeing 737.

As a Reserve Naval Aviator at NAS Atlanta, Georgia, she flew 1,400 hours in the C-9 as a transport aircraft commander with overwater and international qualification.

Lucy is a member of the Ninety-Nines, Tailhook Association, and Women Military Aviators. She is an air safety representative and accident investigator

Fig. 4-8 Cmdr. Lucy Young

for the Air Line Pilots Association, and is a facilitator for USAir crew resource management program. She also holds air transport pilot, flight engineer and certified flight instructor ratings and has over 8,000 hours logged in 40 different aircraft.

In September 1990, Cmdr. Young deployed to Roto, Spain in support of Operation Desert Shield. She made several flights to Saudi Arabia. On her initial visit she found an unwelcome surprise. "The facilities (lavatory) on board the Saudi Naval Air Station were non-existent for females. The females in our crew had to use the aircraft facilities!"

But that wasn't the only cultural shock American women experienced. "The Saudi Naval Operations personnel had not gotten used to female pilots," she said. "So male pilots had to go in and file the flight plan."

Cultural disparities were not the only differences. "The most striking aspect of flying in Saudi Arabia," said Young, "was the lack of contrast in the terrain and the difficulty in spotting the runway. Our perceptive abilities improved, however, once we got used to the 'sand' colors in the Middle East.

"Women played an important role in the Gulf conflict," said Young. "There were numerous women assigned to shore detachments in supervisory roles doing an outstanding job in a very harsh desert environment."

Lt. Lisa Williams - H-46 Helicopter Pilot

Lt. Lisa Williams earned her commission through the NROTC at Duke University, in 1987. She immediately applied for flight training. She immediately ran into a problem too. They told her she was "too small" to fly. "Because of my height, 5 feet 2 inches, and other anthropometric measurements I was considered 'outside the norm.' I was told to forget flying." However, during summer training one year, Lisa met a woman aviator who was exactly her size. "I knew there must be a way around those magic numbers that I didn't fit into." Lisa persevered and got a waiver for her size. But not before they strapped her into certain aircraft and made sure she fit properly. Soon after that "fitting" Lisa got her orders to flight school. See Fig. 4-9.

In January 1989, Lisa won her "wings of gold" and designation as an "Unrestricted Naval Aviator." She then reported to NAS North Island, California to fly the H-46D Sea Knight helicopter.

When Desert Shield began in August 1990, Lisa was with the USS Independence Battle Group and one of the first on station in the Arabian Sea. Her job was vertical replenishment of ammunition, stores, food, cargo, and transportation of passengers and mail.

"We were busy and flew many long flights and racked up the hours as the U.S. readied for Operation Desert Storm.

By January 1991, Lisa was a designated aircraft commander and serving aboard the U.S.S. Flint in the Persian Gulf.

Fig. 4-9 Lt. Lisa Williams

In January 1992, Lt. Williams transferred to NAS Whiting Field to serve as a primary flight instructor in the T-34C aircraft.

Lisa has positive feelings about a life for women in the military. "I believe that military aviation is opening up for women at a decent pace. In my case the helicopter community is not considered a combat assignment so women have exactly the same opportunities as men. We can deploy on ships, fly the same missions, and hold the same jobs. I have always felt like an equal to the other pilots in my squadron."

Lisa also feels that she is judged on her abilities. "I have never felt like I have to 'prove myself' just because I am a woman. I do believe that all pilots have the urge to 'prove themselves' because that's the nature of being an aviator.

"Flight training can be a great career for a woman. But it's tough and the attrition rate is high. Not everyone has what it takes to be a pilot. But perseverance and determination are essential. Learning to fly can seem overwhelming, but the rewards are incredible. The gift of flight is one of the most awesome experiences in life. Whether you are in a single engine, turbo-prop, dual engine helicopter, or Boeing 747, having the earth below, the clouds at your wing and nothing but blue sky in between is indeed a separate peace of its own."

Captain Michelle Tallon - USAF - E-3 Pilot

Michelle's first flight was when she was 16 years old. "I knew I had found my place in life," she said. "I'm not the kind to sit behind a desk. Thanks to my dad I loved working with my hands, and I am mechanically inclined. The two seemed to mesh together perfectly. For me piloting seemed easy."

Michelle graduated with a degree in aviation from Ohio State University in 1983. Her ROTC participation earned her a commission and she applied for flight training. In the early 1980s there was a perceived imbalance of women in military aviation in those areas in which women were allowed to participate. Michelle was not awarded a pilot slot out of college. Her scores were higher than a lot of the men who did receive slots. "I was given a navigator's slot and had to wait two years once I had my navigator's wings before applying for a pilot slot. I received a slot on the first board and went off to pilot training."

While Michelle was waiting for the pilot slot to open up, she flew as a navigator on an E-3 and then upgraded to instructor-navigator accruing 1,500 hours.

After Michelle graduated from flight school she selected and received her first choice aircraft, the E-3. In August 1988, she upgraded to aircraft commander and has accrued more than 2,700 hours of pilot time. Michelle has deployed around the world on "Operation Just Cause," "Desert Shield," "Desert Storm" and "Provide Comfort Operations." Her combined total time as a pilot and navigator is more than 4,200 hours.

"The military has wonderful opportunities for women fliers and you receive the best training possible in the world. Where else can you have access to supersonic aircraft and do things such as landing on a carrier. The doors that are closed to us are rapidly opening and hopefully by early 1993, we will have qualified women flying fighter aircraft designated for combat.

"It hasn't always been easy and it hasn't always been fair but I wouldn't trade the military experience for anything. It has taken me around the world, I've seen numerous foreign countries and have met many new friends. What I have learned 'first hand' could never have been learned through books.

"I love the challenge of the military mission, the military lifestyle but even more, I love the satisfaction of meeting those challenges and doing well. If you are looking for an interesting life - military life is it."

Capt. Terry VandenDolder - C-141 Pilot

Terry began flying in her junior year at the University of Connecticut. She was a full-time student and she found that it took two jobs to pay for flight lessons. That's when she looked into the ROTC.

Terry graduated in 1981 but had to wait a year for a pilot slot in the Air Force to open. Women in the ROTC program must compete on a national level for pilot slots since there are so few openings for women. The men, however, compete at the local ROTC detachment level.

The summer following her graduation, Terry worked as a clerk-typist at NASA in Washington. She spent her lunch hours across the street at the National Air and Space Museum. "I was fascinated by all those who had made their mark in aviation history."

Terry was finally called into the Air Force in May of 1982, but spent the first six months of the year cleaning and fueling planes in exchange for flying time. "I distinctly remember lying on my back on a dolly trying to cut the grease from the bottom of an airplane, using cold water in the dead of winter and thinking, 'there's got to be an easier way to fly.' And there was."

When Terry arrived at Laughlin AFB, Texas, she was the only female in the class of 75 students. "I remember thinking, 'growing up with four brothers didn't prepare me for this.' Although there were a few who were discouraging, many of my classmates were supportive of women military aviators." One of those supporters was a fellow classmate, Glen Lidbeck, whom Terry later married.

In May 1983, Terry graduated from undergraduate pilot training in the top 10 percent of her class. After training in the C-141B at Altus AFB, Oklahoma, she went on to her first assignment as a C-141B transport pilot at Travis AFB, California.

Her first week at Travis saw the crisis in Grenada. While she was not asked to fly; many of the pilots in her squadron were sent to support the airlift mission.

"What left an impression on me about Grenada was that the flight crews that had female crewmembers on them were returned to their bases shortly after the airlift began. Someone in 'higher headquarters' had noticed that there were women on some of the crews and, since the planes were catching ground fire, this could be construed as 'combat.' It was frustrating for me to see that women were asked to do the same job as men until an actual contingency arose, then they were left behind to stay 'safe.'"

Terry flew many classified missions and on one of those missions she had been warned by one of the mission coordinators that she might have to 'relinquish her command' as the aircraft commander to her co-pilot (a male) when they reached their destination, a country where women were not usually in positions of authority. "My co-pilot, (a lieutenant colonel and her squadron's assistant operations officer) responded, 'the hell she will' to that suggestion. When we arrived at our destination, the host military officers initially would not look me in the eye in conversation, but would direct questions regarding the mission to me through the flight engineer. After about five minutes of this triangular game they finally figured out that I was the one in charge. From that point on, we were able to relate one-on-one, professionally."

Terry began noticing many of her peers leaving the service after their initial commitment and continuing their flying careers as civilians with the major airlines. "I was not initially interested in an airline career, but valued the responsibility, leadership and challenges a military career afforded. I became discouraged, however, about the time I attended Squadron Officer School. It was then I first began to realize that the 'glass ceiling' so often referred to by women at mid-level management positions, was also a reality in the Air Force."

Over the following months Terry talked to her senior officers concerning career opportunities for women. Specifically she was looking toward becoming a squadron commander in an operational flying unit - a goal to which most pilots who wish to make the military a career aspire.

"Most of the responses I got were fairly discouraging. Individuals selected for these positions were quite often those who had combat leadership experience. This ruled out women."

Several other factors were operating that focused Terry on what the future held. The aging C-130 and C-141, two aircraft open to women, were to be replaced by the new transport coming off the line, the C-17. And during this period, it had not been officially opened to women. In addition, the Air Force was restructuring its pilot training program in a way that would possibility exclude women from flying and instructing in the T-38, the only high performance jet aircraft in the Air Force inventory open to women. "Instead of doors opening, I saw them closing."

In view of these mounting obstacles, Terry left the active service in September 1989 and joined the Air Force Reserves. In January 1990 American Airlines hired her as a flight engineer on the Boeing 727.

In August 1990, when Saddam Hussein invaded Kuwait, Terry and her reserve unit were activated. Their mission at first was to fly cargo and troops to Saudi Arabia, Bahrain, United Arab Emerits, etc. "The gradual escalation of our involvement could be recognized by what we were carrying in the back of our plane. We started carrying cargo necessary of the set up of operations, like electronic equipment, trucks and generators. Then when the ships were deployed that could carry larger quantities of heavy equipment we began carrying items of a more 'timely nature' like mail for the hundreds of thousands of troops that were already deployed to the region. In November we carried turkeys for their desert feasts and as we drew closer to the deadline in January we began airlifting troops by the thousands from U.S. bases in Germany. In January we brought in tons of armaments that would be later used by the fighter aircraft.

"On the night of January 15, my crew was flying from Germany to Dhahran, with the airplane filled from top-to-bottom, front-to-back with empty metal caskets. I remember staring at those caskets wondering how we would bring them back. Later I flew 21 KIA's (Killed in Action) from Saudi Arabia to Frankfurt, Germany, on the first leg of a solemn trip back to the United States. Some were casualties of the scud attack on the barracks at Dhahran. Three of these were women.

"What we saw in Desert Storm was an American public that was initially uncertain about women being deployed to a war zone. At first some military commanders were cautious about using women in positions for which they were trained. But when push came to shove, a professional get-the-job-done attitude prevailed. Women were incorporated fully into the operations. Military leaders knew that women were highly capable of doing their job and they performed very well."

Capt. VandenDolder sees progress since the Grenada incident when women were returned to base to avoid the risk of combat fire. In Desert Storm more than 32,000 women served in key combat support missions. They stood side-by-side with their male counterparts and flew helicopter, air refueling and reconnaissance aircraft. They have proven their invaluable worth in their contributions to the successful outcome of the war.

"If you wish a career in military aviation," Terry advises, "Go for it. It's one of the best jobs, both on the planet and off. It means hard work, and it takes a lot of determination but the rewards are well worth the effort. Flying can put a new perspective on one's life, both literally and figuratively. You can gain from this unique perspective a new appreciation, as I have, for nature, the laws of physics and the art of a good landing.

"The sky is the limit is an axiom to which we can all aspire, but only so long as we refuse to let artificial laws, restrictions and cultural mores deny us our basic right to equal opportunity for equal skill. Women have proven they have the skill, cunning and ability to conquer any task. Being qualified for a position is the product of the talents and determination of the individual, and is not prescribed by gender."

Col. Kelly Hamilton USAF KC-135 Pilot

Col. Kelly Hamilton took her first flight in a Cherokee 140B in 1969. Today she still flies civilian aircraft in her spare time but her machine of choice is an Air Force KC-135 air refueler.

Kelly earned her commission as a Second Lieutenant through Officer Candidate School in 1973, before the Air Force opened flight school to women. Her job at the time, however, did get her into the cockpit as an avionics test and evaluation officer. In this capacity she rode the back seat of such sophisticated machines as the F-4, F-5, F-111, T-38 and the A-37.

When the Air Force opened pilot training to women Kelly signed up. She was among the first twenty selected but the Air Force divided the group and she graduated from the second class. On the day she graduated she received her father's original silver wings, and another set of wings from Col. Michael Brazelton, a six year P.O.W.

Following graduation Kelly received permission from the commander of the Tactical Air Command to fly an air-to-air D.A.C.T. (Dissimilar Air Combat Training) in the F-5A Aggressor. This flight made her the first woman pilot in the Air Force to fly a fighter aircraft since the WASPs of World War II. This marked the first of many ground-breaking firsts that Kelly Hamilton would achieve.

In 1978, Kelly graduated from SAC Combat Crew Training as a KC-135 co-pilot in the top ten percent of her class. She would repeat her excellent performance when she went through the class two years later for pilot training and this time she would add to her record the #1 graduate.

Kelly then went on to become the first woman to land a military aircraft in the United Kingdom since World War II.

By 1982, Capt. Kelly Hamilton led the first all-female flight crew in the Air Force. This historic refueling flight was dubbed "Fair Force One."

In 1985, Kelly accepted an assignment as Squadron Commander, Cadet Squadron 28 at the United States Air Force Academy. She was responsible at this time for 120 cadets and also served as a flight instructor. It was here at the Academy that Kelly began her work to help repeal the USC Title 10 Combat Exclusion Law. She served an active role on the Defense Action Committee on Women in the Services (DACOWITS), a presidential appointed committee that began working toward getting the law repealed that precluded women from assignment to combat aircraft.

Kelly's dedication to this goal earned her election as president of Women Military Aviators, an international organization, in 1988. In 1989, she was reelected to a second term.

By 1990, Kelly Hamilton's career had led her to become Director of Flight Training for the largest air refueling wing in the world.

Like many Americans Kelly saw the United States headed for war in the Middle East. Her expectations were that the women were trained professionals and would deploy like the men. She did not see this happening and took a bold step. She challenged the senior commanders on their decision not to deploy women to Desert Shield. As a result of this one-on-one confrontation, the policy was changed and women were deployed along with the men. This was the first crack in the years-old armor of male protectiveness.

Col. Kelly Hamilton flew over 200 hours during Desert Shield/Desert Storm. Her missions included 29-ship formations that needed fighter-aircraft cover. For her efforts she won the Aerial Achievement Medal, but her hours were not considered "combat hours." Only fighters, bombers and attack aircraft may log "combat" time. The Air Force definition of combat is the ability to shoot at, or inflict damage on the enemy. At the time she commented, "It has nothing to do with being at risk. Hopefully the American public will see the hi-tech aspects of war and recognize that brains and agility are the key elements to flying "combat" aircraft and not brawn."

The American public did recognize the contributions of women like Kelly Hamilton and the years-old debate about the role of women in combat took on a new intensity. The women had flown the missions and the facts were stronger than the protective rhetoric spewing out of Congress.

After Desert Storm, Kelly was assigned to the Joint Staff USCINCPAC Commander-in-Chief Pacific. She continued to work through DACOWITS and Congress to successfully repeal the combat exclusion law. She personally wrote more than 50 letters a week on this subject to legislators. Her efforts and those of many others paid off on July 31, 1991, when the Roth-Kennedy Amendment passed. When the presidential commission was formed, Kelly Hamilton was there to testify as to the capabilities and readiness of women to fly combat aircraft.

Kelly Hamilton defines the bottom line of the issue this way, "My profession is the defense of our nation and its ideals. I've spent all my adult life being told by people in the system that you can't do that, you're a woman. I have always taken great pleasure in proving them wrong. When it comes to flying, the airplane does not know my gender, it only respects my talent as an aviator."

Cmdr. Carole Danis Litten - USNR

Scores of women who qualified to serve in Desert Storm were not deployed. They waited with anticipation to serve their country and to preform the duties they had trained for throughout their professional military careers, but not everyone did get the call. Many, like Naval Reserve Cmdr. Carole Danis Litten, even volunteered to go, but as one Pentagon official remarked to her, 'What do you think you're going to find in Saudia Arabia? Submarines in the sand?'

"Units didn't go over as a whole," explains Carole, "Often times only specialized individuals are activated, which is common during deployment. It was frustrating because I expected to go, and even volunteered, but they didn't need all the officers." See Fig. 4-10.

Carole joined the U.S. Navy through the officer candidate program in Newport, Rhode Island in January 1977. As a first for the Naval history books, she was commissioned by her grandfather, Admiral Anthony L. Danis Sr., USN and by her father, Cmdr. Anthony L. Danis Jr., USN. After a tour as operations watch officer at the Naval facility at Grand Turk, in the British West Indies, Carole sought to follow the family tradition as a third generation Navy pilot. She graduated from the third group of women admitted into the Navy's flight school and was selected for P-3 ASW (anti-submarine warfare) training at Jacksonville, Florida.

Carole reported to VXN-8 (oceanographic research) at NAS Patuxent River, MD, where she completed a three year tour involving worldwide missions that operated out of more than 25 different countries, including the Arctic and Antarctic. "My first tour as an aircraft commander was in Brazil and the mission required that we wear civilian clothes. Those who saw me emerge from the plane with my male crew members watched as we were escorted to the waiting embassy vehicle. They were convinced I was a prostitute who had been flown in for a classified mission of a different nature!"

Carole reported to VT-6 NAS Whiting Field, Florida, as a flight instructor and transferred to the reserves in 1986, after more than ten years of dedicated service to the Navy. Within the Naval system Carole was an officer and a leader who enjoyed a successful career. But the core environment was not free of sexual harassment and Carole grew tired of being treated less than equal among the men.

She turned towards the airlines for employment and actively pursued two carriers, Delta and Piedmont. Though she had more heavy flight time and experience than the other men leaving her squadron to interview with Delta, it was Piedmont that invited Carole for an interview. She successfully passed each applicant phase and Piedmont hired her on September 25, 1986.

Carole qualified first as a B727 flight engineer. The day was June 27, 1987 and Carole was a member of the crew aboard the airline's first all-female B727 flight. She completed her probationary year by upgrading as a first officer on the B737-200 and flies today as a co-pilot on the B-737 300/400. She is an active member of the International Society of Women Airline Pilots and serves as one of the organizations Executive Council members. Her active participation and professional presence on airline panels and other projects have helped make women airline pilots a more visible force within the industry.

As a reserve officer, Carole has served with the Reserve Patrol Wing (P-3) Atlantic, C-9 squadron and on surface tender ship. She is currently an executive officer of the Naval Reserve Air Station in Keflavik, Iceland, and drills at the Naval Air Facility at Andrews Air Force Base in Washington, D.C.

In 1992, Carole was selected by the Office of the Assistant Secretary of the Navy as the only female aviator to assist in the development of an inclusive sexual harassment training package. The stand down was ordered in the wake of a sexual harassment scandal that erupted when 26 women Navy personnel were assaulted during the 1991 Tailhook convention. The incident forced the Navy to finally focus its attention on sexual harassment issues within its core. The Navy has as a result, implemented policies and procedures that aim to drive the unacceptable behavior from its ranks.

Having served as a crewmember, and an aircraft commander exposed to a wide range of sexism, Carole's role on the review board was crucial. "The Navy

Fig. 4-10 Cmdr. Carole Danis Litten

recognizes the undermining current that sexism has on the core values of integrity, professionalism and human dignity. The training package addresses the immediate legality of sexual assault and the punishment assigned to such behavior. It educates personnel on just what constitutes unacceptable behavior and strives to cultivate an environment where the rights, concerns and contributions of all Navy personnel are recognized and respected."

Carole is encouraged by the steps the Navy is taking, their commitment to the program, and their "zero-tolerance" platform, but believes that in the end it is individual "attitudes" that must be changed. "The women in combat issue is a perfect example of how superior attitudes are fostered among the men. This ultimately breeds the very sexism and harassment we are trying to eliminate. The division of gender in combat results in the belief that women's duties are inferior and segregates the women as second class citizens. Women need to be allowed into combat roles and to work with the men as equals. Only then will they gain the respect required to eliminate tones of discrimination and harassment all together."

Carole is currently addressing these issues as she provides feedback to a presidential task force on the assignment of *Women in the Armed Forces.* "The controversy regarding women in combat is certainly not new. Finally, all fronts of the military as well as the House and the Senate acknowledge that women can effectively preform within the lines of battle. The real task now is to culturally break away from the limits society has imposed on us and to change *their* 'attitudes' on how women will be viewed in combat."

Aftermath

As a result of the important role played by women pilots in the gulf conflict, Congress reconsidered the need to eliminate the decades-old law prohibiting women from flying aircraft engaged in combat missions. The review had come as a result of women flying non-combat support roles in the Persian Gulf War that took many of them behind enemy lines and exposed them to hostile fire. Army helicopter pilot Major Marie Rossi had also become the first woman aviator to die in the line of duty. The cease-fire had just begun when Rossi, blinded by bad weather, crashed into a microwave tower.

In mid June 1991, after the smoke had cleared from the battlefield, Congress voted to lift the federal restrictions on women pilots flying in combat. The bill went to the Senate where it met stiff opposition. On July 31, 1991, to the surprise of a well-organized opposition, the Senate voted overwhelmingly to pass an amendment to the Defense Authorization Bill that rescinded the "combat exclusion" in place since 1948. Through their commitment and performance during the Gulf War, the women had proven to die-hard skeptics that "The Right Stuff" has no gender. In spite of this, the reality of women flying combat aircraft in a combat environment remains controversial.

And the debate goes on, still centering around women's psychological and physical tolerance. Can women tolerate the physical demands? Can women fly and operate weapons under G forces eight and nine times as strong as gravity? Because women pilots make up less than 2 percent of the Navy and Air Force's 32,000 pilots, little specific research has been done to answer the question. But this is not a new issue.

Dr. Kent Gillingham, an aeromedical researcher at Brooks Air Force Base says, "If a person doesn't have the right degree of G tolerance, he or she is going to be too busy staying awake and alive to deliver weapons."

Gillingham's research shows that women subjected to as much as 7Gs had essentially the same tolerance as men. "Any differences linked to sex were offset by women's shorter height and by the correspondingly shorter distance from their hearts to their heads." (The shorter distance enables blood to reach a woman's brain faster than a man's.)[10]

In the newer jets like the F-16, a pilot experiences forces of +8 or even +9Gs. "We don't have any definitive data," Gillingham says, "but we do know that some women can tolerate such forces." His conclusion: "There is no reason categorically to exclude women from flying on the basis of G tolerance."[11]

Cmdr. Rosemary Mariner,[12] who is a graduate of the first women's class to go through Navy flight training has listened to these debates for almost 20 years. She points out that women have flown for years in the NASA shuttle program. In the Navy at least, says Mariner, women pilots fulfill the same requirements as the men, G tolerances included. Women today routinely train the men in combat techniques in the aircraft that fly combat. Studies have shown that some of these women who train fighter pilots handle high-G conditions better than men.[13] "The most important thing is the pilot's brain, heart and mind, not the package they come in," says Mariner.

Another Navy pilot agrees. "The real battle is going to be mental. I think women are more than capable of holding their own in that arena."

Ostensibly, the resistance to expanding roles for women in military aviation, including women flying military fighters, has been based on physical strength and toilet accommodations. But the issue is really power and control, not physical strength. The real power and control, most military officers will agree, is in the combat roles. Until women can assume combat roles, they will have no real power. The structure of the military fighter pilot world will not make their transition easy either. The fighter pilot community is insular and macho by nature. The men look at themselves as an elite group and prefer to keep it that way. Why open the avenue to others if you don't have to? And if it becomes a mandate, why make it easy? These sentiments seem to be common feelings expressed among fighter pilots.

At least one Air Force general believes women can do the job. "Women can fly fighters, and women can pull the Gs," said Lieutenant General Thomas J. Hickey. Navy Vice Admiral Richard Dunleavy also agrees. In response to male combat pilots who say they won't accept women combat pilots, Dunleavy says,"Poppycock, women can pull Gs just as hard as men."[14] Others disagree. In July, 1991, Congress heard Marine Corps General Robert H. Barrow say, "I may be old fashioned, but I think the very nature of women disqualifies them from combat. Women give life . . . they don't take it."

This is exactly the type of thinking that is stonewalling advancement for Navy and Air Force women. Without combat experience, women are denied job opportunities and promotions open to men. In effect, they are denied the use of their full potential.

Today in the Air Force, 97 percent of all positions are open to women. There were 3,800 women deployed to the Persian Gulf and Air Force women pilots flew and crewed strategic transport, tactical transport, tankers reconnaissance (AWACS), and aeromedical airlift aircraft.[15]

In the Navy, currently 59 percent of job positions are open to women. Women cannot serve on ships that are expected to be engaged in combat but can be assigned to ships of the Combat Logistics Force. However, women are allowed to serve temporary duty on combat ships as well as train men to fly combat planes.[16]

In the Marine Corps, only 20 percent of job positions are open to women. The lower figure is partially explained because the Corps has no medical or chaplin branches of its own; it receives these services from the Navy. About a thousand women Marines deployed to the Persian Gulf, but none as pilots.[17]

In the Army, 52 percent of the job positions are open to women. About 26,000 Army women deployed to the Persian Gulf. Women participated in the initial invasion into Kuwait and Iraq. They were assigned to forward support units in various specialties including flying helicopter, to transport personnel, equipment and supplies. Two women commanded battalions - a military police battalion and a material maintenance, and women were also in command of companies, aircraft squadrons and platoons and squads in a variety of units. Eleven Army women lost their lives in Saudi Arabia. Five of the 122 U.S. troops killed in action were Army enlisted women, and one a helicopter pilot. Two Army women, a truck driver and a flight surgeon, were among the 25 United States personnel held prisoner of war by Iraq.

Although the Coast Guard is a uniformed service, they are part of the Department of Transportation, and the statutory restrictions that pertain to military women in the Department of Defense do not apply to Coast Guard women. Thirteen Coast Guard women deployed to the Persian Gulf, serving in port security positions.

The one shortcoming of the Congressional decision to remove the combat exclusion was they empowered the service secretaries to employ women aviators in this new role, at their discretion, without legal barriers. Congress, as a compromise to the opposition, recommended further study of the issue.

In March 1992 the President of the United States authorized, as part of the national Defense Authorization Act the formation of a presidential Commission to access the laws and policies governing the assignment of women in the military. Fifteen commissioners were appointed by the President. Some of the issues they are exploring have to do with the effect of pregnancy on readiness and whether changes in the combat exclusion laws would subject women to a military draft.

After the commission began studying the issue, an article appeared in the *Washington Times*. According to the writer, Rowan Scarborough, it would be at least a year before a decision on letting women compete for combat pilot jobs was made. According to the article, Defense Secretary Dick Cheney will not order any change until after the report is filed by the commission. The report was not due to be published until December 15, 1992.

Senator William Roth, Jr., a Delaware Republican who sponsored the amendment striking the combat ban, had termed the commission idea "hogwash." He is quoted as having said, "I think the commission should not be used as an obstacle to women combat pilots." He went on to say, "The women pilots in Desert Storm had proven their skills and he reportedly said, "I tell you, we ought to have a gender-neutral policy as respect to pilots."

Chapter Five

Sport Aviation

It is not easy to be the best. You must have the courage to bear pain, disappointment, and heartbreak. You must learn how to face danger and understand fear, yet not be afraid. You establish your goal, and no matter what deters you along the way, in your every waking moment you must say to yourself, "I can do it." Betty Skelton

Air shows have been around since the early 1900s and today in the United States there are almost 400 professional, colorful and entertaining shows held every year. Air shows are also the second largest spectator sport in the United States. Baseball draws twice as many people as the National Football League and auto racing combined, but the 22 million annual spectators who attend air shows every year is testimony to its extreme popularity.[1] Among the 210 professional performers on the air show circuit, aerobatic pilots get high visibility. Aerobatic flying takes countless hours to perfect. It requires constant access to an aircraft, and the financial resources to pay for the costs associated with the flying and maintenance of the machine. It is also one profession where only the best remain.

Betty Skelton - First Lady of Firsts

In aviation circles there are few people who are considered "living legends." Legends are those who have blazed trails and whose glorious exploits, impressive accomplishments and immeasurable popularity has spanned generations.

In an era where heros were race pilots, jet jocks and movie stars, Betty Skelton was an aviation sweetheart, an international celebrity and a flying sensation. Her career and success could be pages right out of a storybook but even Hollywood couldn't produce a picture as grand as the real life that Betty Skelton has led.

Betty's enviable record is still recognized today by pilots and competitors, and she is frequently referred to as, *The First Lady of Firsts*. She was the first woman to cut a ribbon while flying inverted and she piloted the smallest plane

Fig. 5-1 Betty Skelton

ever to cross the Irish Sea. Twice she set the world light-plane altitude record (29,050 feet in a Piper Cub, in 1951). The first time Betty broke the record established in 1913 in Germany. She also unofficially set the world speed record for engine aircraft in 1949. Today Betty Skelton holds more combined aviation and automotive records that anyone in history. "I have always been interested in speed," she said. "It's pretty fortunate when you can find something you love to do so much and it is also your occupation." Besides winning the International Feminine Aerobatic Championship for three consecutive years, 1948 - 1950, she received honorary wings from the United States Navy and held the rank of Major in the Civil Air Patrol. She flew helicopters, jets, blimps and gliders, and participated in all U.S. major air events in the forties. Today the very spirit of Betty Skelton and her love and devotion to aviation and aerobatics is bestowed upon the top female in the field with the presentation of the coveted "Betty Skelton First Lady of Aerobatics" trophy.

As a young girl in Pensacola, Florida, Betty watched the Navy Stearmans fly their aerobatic routines from her back yard. She dreamed of someday flying too. Her parents co-owned a fixed base operation in Tampa, Florida so she literally grew up in airplanes. Betty recalls playing with model airplanes instead of dolls and soloed illegally at the age of twelve. She was a commercial pilot at eighteen and a flight instructor by the time she was twenty.

Betty's real desire was to fly military planes as a ferry pilot in the WASP program. Her flight experience ranked among the best but she was "shot down" because of her age. The minimum was 18 1/2 and at 17 she couldn't get a waiver. The WASP program would end four months before she reached the appropriate age. In the years while she waited, Clem Whittenbeck, the famous aerobatic pilot of the 1930's, taught Betty how to do loops and rolls and for her, aerobatics was love at first flight.

"My first aerobatic plane was a real crate!" says Betty, "It was a 1929 Great Lakes that was sluggish and not nearly as responsive as a true aerobatic plane should be." Its Kinner engine was usually "sick" and she had her share of serious mishaps and narrow escapes flying the old "Lakes." She even crashed it once.

In 1948 she purchased the one and only Pitts Special, an experimental biplane hand-built, and carefully designed by Curtis Pitts himself. It was a single seat, open cockpit aircraft weighing only 544 pounds. Little did she know at the time that she and the Pitts would become a famous team that would soar to new heights. Her plane would became the most famous aerobatic aircraft in the world. Betty would teach the little Pitts a lot of new maneuvers, but not before the airplane taught Betty a thing or two. See Fig. 5-1.

With one Feminine Aerobatic Trophy already to her credit, Betty flew her new Pitts Special back to her home base at Tampa's Peter O'Knight airport. A large crowd of friends and well-wishers gathered to welcome Betty and her new flying machine, and it was on that day she found a suitable name for her new partner.

Upon purchasing the Pitts, the former owner quickly checked Betty out in the plane but neglected to mention a few things about the landing characteristics. Betty made her approach into Tampa and landed nicely. Suddenly, as she brought the Pitts to a stop, it went out of control. She could not have known that at low speeds the plane's rudder was ineffective and only the brakes would keep her going straight. As she clamored to bring the plane under control she said under her breath: "You little stinker!" The Pitts resented that rejoinder and ground-looped in front of the entire crowd. Nothing but Betty's pride was hurt but the name stuck. From that moment on they would live forever as Betty Skelton and her *Little Stinker*.

Betty would practice for hours, sometimes on just one maneuver. She disciplined herself and never strayed from the hard and fast rules that she imposed. One rule was to build an extra margin of safety into all her low level maneuvers. She would compromise for nothing less than an extra ten percent of airspeed or altitude. If she had both, so much the better. This extra margin of safety would keep her out of trouble in the future. With Betty safety always came first. She routinely practiced at an altitude of 3,000 feet unless she was practicing her ribbon cutting, in which case she always had a spot picked out below in case she ran into trouble and needed to make a forced landing.

Fig. 5-2 Betty Skelton and "Little Stinker"

Betty had two safety belts built into the airplane at different locations in the event one broke loose during a violent maneuver. During practice she placed towels between herself and the safety harness to act like shock absorbers, but they never eliminated all the pressure. As a result, Betty was always bruised. When she worked intensely on her outside maneuvers, the forced pressure of blood to her face would cause Betty to go for weeks with black eyes and splotches on her face. During other maneuvers when the blood would drain from her body, she would experience momentary "red-out." She slowly built up her tolerance and always knew how far she could push a maneuver and still stay in control.

Soon Betty and *Little Stinker* worked as one. Her movements telegraphed calm and confidence and her little friend, listening to her thoughts, reacted as an extension of Betty's body, diving and looping to her every command.

The maneuver Betty is best known for is the inverted ribbon cut. This involves cutting a ribbon strung between two poles, ten feet from the ground, upside down. You could count the number of men on one hand that could do the stunt, and being the best meant that Betty had to learn how too. Her friends tried to talk her out of attempting the stunt because it was a dangerous maneuver that allowed for no margin of error, but Betty was determined. She would be successful at becoming the first woman to complete the feat but with near fatal results. See Fig. 5-2.

Betty promised her friends she would attempt the stunt scientifically, making many practice runs until she felt comfortable. She would start with a high altitude, line up the poles, roll upside down and go lower and lower until she got a feel for it. Betty's first pass was deliberately high but right on target. Feeling confident, she set up to cut the ribbon on her second attempt, only on this try the Pitts' engine quit just as she rolled upside down. She was only a few feet above the pavement.

Betty's ten percent rule saved her life that day. The extra ten percent in airspeed enabled her to execute a half outside snap-roll and bring the plane and herself down to safety. An inspection of the fuel injection system found that the engine's injector jets were clogged with dirt. After a simple cleaning her *Little Stinker* was as good as new.

Being number one meant Betty had to work twice as hard as everyone else. Perfecting the artistry of aerobatics and taking her routines to the level of flawless perfection was a skill Betty worked at constantly. Betty attributes her success in the sport to having had the coolness of a champion bullfighter, the fighting spirit of a cornered cobra and the dedication of a priest.[2]

One of the hardest things about flying the aerobatic circuit was enduring the pain of watching many friends die. Each fatality occurred while trying to out fly Betty or while trying to imitate maneuvers that only she could do in her Pitts. (Death was a constant reminder that all aerobatic pilots fly in the shadow of death.) "At times it felt like a waiting game," says Betty, "wondering who would be next." Betty looked on death as one of the risks of the business. "Learning to fear death without actually being afraid was something you had to do to make it through."

Betty didn't just "luck" into her accomplishments and achievements because of her good looks, although she was every bit the glamour gal that the press made her out to be. Betty didn't believe in luck, good or bad, just great timing. Betty has, however, become partial to the number "2" over the years.

Betty first laid eyes on her little Pitts on the 2nd of the month. It's serial number is #2 and the aircraft registration number was 22E, (a number that is still reserved for Betty today.) Even the old Taylorcraft she first soloed was NC 22203 and her Great Lakes was numbered NX 202K. Being born at 2:00 a.m. and wearing Channel No. 22 perfume is probably just a coincidence, but number "2" would have special meaning to Betty as the years and her career flew by.

Besides setting altitude records and flying aerobatics, Betty felt the need for speed and loved to race. "Some of my fondest memories are of the flying at the Cleveland Air Races in the late forties." Betty remembers, "Every big name pilot in the country would be at the races, all the old gang, and it was easy to get caught up in all the excitement."

Being a champion meant Betty was always in demand and maintaining the hectic schedules was physically exhausting and emotionally draining. Air shows and appearances were scheduled one after another, sometimes three in the same day and all in different cities. Betty's life was anything but normal and she tired of her nomadic lifestyle. She realized early that it would be impossible to handle a career as well as a family. "During those years more than one engagement ring was returned."

With all the accomplishments Betty had made in aerobatics, there didn't seem much left for her to do. For all the fame and glory, there was little money to be made in the sport Betty loved so much. It was no secret, flying was an expensive undertaking and there weren't many alternatives for women in aviation. She had been born too late to qualify for the military WASP program and too early to try for the airlines. Staying in aviation would have meant going back as a flight instructor or flying at a fixed base operation, "hardly a suitable challenge." So on October 2, 1951, at the age of 26, Betty retired from professional aerobatic flying.

Although Betty would leave her first love of flying, she would be instrumental in forever changing how women were viewed in the competition level of aerobatics. She challenged the Professional Race Pilots Association and made changes that today allow women to race in closed course pylon races with the men. This set up a chain reaction that would open many doors to women at the competition level of aerobatics.

As for her *Little Stinker*, it was enshrined in the Smithsonian's Air and Space Museum on August 22, 1985. "It is tremendously gratifying to know that children may see the tiny plane for decades to come and hopefully be inspired and seek a future in the sky and space."

Being a person who thrived on challenges and with the open road ahead, it seemed only natural that Betty moved on to auto racing. Her records in this field would become as impressive as those in her aviation career. Betty broke the world land speed record for women four times, and became the first woman to officially drive a vehicle over 300 mph, (315.72 to be exact)! She won nine sport car records for speed and acceleration, and became the first woman test driver as well as the first woman to drive an Indianapolis race car. She toured for the National Safety Council, appeared in national advertisements and TV commercials and co-wrote and produced an award-winning motion picture movie entitled, "Challenge."

Betty was inducted into the Florida Hall of Fame in 1977, The International Automotive Hall of Fame in 1983, The International Aerobatic Hall of Fame in 1988 and The Tampa Bay Walk of Fame in 1991. Betty married Donald Frankman, a TV director/producer, on New Year's Eve in 1965. They make their home in Florida where they own a real estate business and consider themselves semi-retired. As a woman who has set more combined aviation and automotive records than anyone in history, Betty can still be found flying in her airplane or driving her red Jaguar, but now it's purely for pleasure.

When Betty decided to fly *Little Stinker* she shared with her little friend her beliefs. "If you are to fly with me you must believe in yourself. Pay no attention to what others say, what they think, what they do. Let your free spirit take you where you will, and when it falters, let your soul demand that you not give up, but only aspire to climb higher and higher.

"As you soar into the heights, never forget the others or look down upon them. Remember, you were once there and needed help, understanding and love. Direct your free spirit to helping your fellow man...and you will know heaven on Earth.

"Never believe your own press, nor take your accolades too seriously. It matters not what you did yesterday. Only what you do today and tomorrow is meaningful to your freedom and spirit.

"Challenge and perfection is the greatest gift of life. Embrace it and use it well. To turn your back on the challenge of perfection is to close the door on your spirit, your freedom... your very existence.

"It is not easy to be the best. You must have the courage to bear pain, disappointment, and heartbreak. Our dedication must help lift the other up when one of us is down. You must learn how to face danger and understand fear, yet not be afraid. You must establish your goal, and no matter what deters you along the way, in your every waking moment you must say to yourself, 'I can do it.'"[3]

Patty Wagstaff - World Aerobatic Champion

There are few sports that demand the physical and mental discipline that it takes to be an aerobatic pilot. That is probably why competitive aerobatics has become a popular sensation for pilot thrill seekers and spectators.

Performing at the competition level requires that the pilot have split second timing and precise speed and altitude control. The critical factors of wind and temperature must constantly be calculated and only the most advanced aircraft can execute the maneuvers. A wrong turn, a misjudged pullout or a roll in the wrong direction will not only cost the pilot points, but could cost them their lives.

Until 1972, men and women competed separately in aerobatic competition. For decades, women excelled in this precision sport with skill and artistry but none had captured the overall National Aerobatic Championship title, at least not until Patty Wagstaff streaked into the competition. See Fig. 5-3.

"My father was a B-747 captain for Japan Air Lines and we lived abroad for many years. I spent my early years in Japan but attended schools in Europe," said Patty. She continued her education in Australia before moving to Alaska in 1978. "There were no air shows to inspire me when I was growing up in Japan but I remember flying with my dad as a kid in a DC-6 and DC-7." Patty always knew she wanted to fly, but "there just wasn't a career in it for women back in those days," remembers Patty.

Aerobatics always interested Patty, but it wasn't until her husband Bob taught her the basic aerobatic maneuvers, that she was hooked. "To me there was no challenge to straight and level flying, and the task of flying from point A to point B seemed very boring." It was the thrill and rush she felt when doing aerobatics that put her on the road to competition flying.

Fig. 5-3 Patty Wagstaff

"I entered my first competition in 1984 and made the U.S. Aerobatic Team in 1985. I set my goals very early and let it be known that I wanted to win the best overall U.S. National Aerobatics Championship title." This was an unprecedented statement. Many people were skeptical "Even the women I competed with didn't think a female would win the title." "Judging is always subjective," Patty explains, "but soon I began to feel as though I was being judged as a pilot and not a 'woman' pilot. Most of the judges on the circuit are men and I think they learned that I was a determined competitor." Beginning in 1987, Patty was the top-ranked U.S. woman aerobatic pilot and had captured many first place titles among the women. She remained steadfast in her quest to become the best overall champ and soon men and women alike rallied to her support. They began to believe that if a woman could win the title, it might be her.

Patty's road to victory has been paved with many achievements. She received the Harold Neumann Trophy of Aerobatic Achievement in 1986 and the Rolly Cole Memorial Award for Contributions to Sport Aerobatics in 1987. She also won the coveted Betty Skelton "First Lady of Aerobatics" Trophy in 1988, 1989 and 1991. There was no doubt she was becoming a major contender for the best overall national title.

Flying aerobatics at the competition level requires hours of practice and training. "My home in Anchorage, Alaska doesn't provide the most favorable weather so when I prepare for the competitions I head to my training base in Tucson, AZ." There her sleek and sporty aircraft awaits her. Together they spend 2 to 3 hours a day in the practice area of the Arizona desert, creating new routines, sharpening skills and perfecting maneuvers.

Patty's state-of-the-art, German-built Extra 260, is her 5th flying partner since she began competing, and her favorite. "It's a single seat monoplane made of wood for light weight performance, and designed exclusively for professional aerobatics." It has a 300 horsepower engine that allows Patty to climb at 4,000 feet per minute and has a roll rate of 360 degrees per second. "I practice a wide range of maneuvers during my sessions. From vertical snap rolls to lomcevaks, (lomcevak means "headache" in Czechoslovakian,) torque rolls and tailslides, hammerheads and spins. Snap rolls and rolling turns are my favorite because they're fast and fun." See Fig. 5-4.

Although Patty wears a parachute in all her performances, she has never had to use it. "I saw one of my friends 'bail-out' from a maneuver that had gone wrong and that convinced me that I needed to face my fear and learn how to parachute."

Sherman, Texas hosted the 1991 U.S. National Aerobatic Championships and Patty was ready. She knew this could be her chance to best the competition. "Flying is the easy part," Patty says, "It's the pressure that can make the difference between winning and losing a competition." There are four main categories: Sportsman, Intermediate, Advanced and the Unlimited. Patty competes at the most challenging level, the unlimited category. Pilots in this category fly the most complex and difficult set of maneuvers of all the competition levels.[4]

First, Patty had to complete the Known Compulsory program which includes basic maneuvers. The Free Style event was next. It allowed Patty to perform her personalized program. The four minute routine is usually the most exciting and entertaining to watch. The most difficult portion of the competition is the Unknown Program. This sequence of maneuvers is handed to the pilot only hours before her flight and practicing the routine is not allowed.

With approximately eighty entries, and thirty in her category, Patty flew a perfect performance. "As long as I stayed on top, no one could catch me." She accumulated the most points over both the men and the women and achieved her goal of winning the U.S. National Aerobatic Championship.

Becoming the first woman national aerobatic champ is a true milestone for women in aviation. The publicity has brought her into the limelight and she is in popular demand for speaking engagements, public appearance and air shows. No one knows the business better than Patty. She has become a professional business woman and an aerobatic competitor. She takes time out to act as her

Fig. 5-4 Patty Wagstaff

own manager by scheduling and arranging all her own appearances. "Ironically all this takes me farther away from what I really love doing most, flying!"

"Air shows are pure fun for me and they are the perfect contrast to competition flying." Patty is booked throughout the year giving air shows all over the country. Performing her dazzling shows not only thrills her fans but allows her to keep her "G" tolerance when she's not competing. "I choreograph all my routines and design them for each location and event, because safety is most important." Her maneuvers are styled and executed to show the ultimate in precision flying and her heart-stopping low level shows hold her audiences in suspense. "I really enjoy the reaction from the crowds and love meeting the people to sign autographs and answer questions after the shows.

"Because of the title, people think I am someone exceptional, but I really don't think that I am. I have to practice just like everyone else." As for the future,

Patty looks forward to representing the U.S. in Olympic-level international competition at the 1992 World Aerobatic Championships in Le Harve, France and she'll defend her current title in the next national competition.

Patty speaks fondly of her sister, Toni Combs, who also chose a career in aviation. "Hers is an area of expertise that falls at the opposite end of the flying spectrum." Airline flying. No snap rolls or inverted flying for her B-727 pilot sister who flies for Continental Airlines, based in Guam. "Patty and I are good at what we do and we don't feel we have to compete with one another in any way." says Toni. "I'd rather be right side up with a crew meal in my lap and she'd rather be upside down. I took my first aerobatic lesson with my famous sister in Pompano Beach, FL, and felt completely secure with her at the controls. When I'm in the States I love going to the air shows to watch Patty perform. I especially like watching the crowd's reaction which makes me want to shout, "Hey, that's my sister up there!"

Patty shares her aerobatic passion with her husband, who has provided support and encouragement throughout her career. Bob is an attorney in Anchorage, and the president of the U.S. Aerobatics Foundation. "We both judge events that I'm not competing in and we design maneuver sequences for different levels of aerobatic competition." A world turned upside down is a way of life for the Wagstaffs, and that's the way they like it.

Dorothy Hester Stenzel - Aerobatic Pilot

Dorothy Hester Stenzel learned to fly in 1927, at the age of 17, and worked her way through flight school by parachuting at local air shows for money. By 21, she was barnstorming and performing as a pilot with Tex Rankin's Flying Circus, billed as "Miss Kick-a-Hole-in-the-Sky." Dorothy's flare for outside loops and snap rolls won her acclaim and reserved her a place in the aerobatic history books for generations.

On May 15, 1931 Dorothy set a record of 56 inverted snap rolls, a record that has yet to be broken by any other man or woman. Two days later at an air show in Omaha, Nebraska she established an unbelievable record of 69 consecutive outside loops (seven of which were not officially counted because they were not judged to be perfectly shaped.)[5]

The outside loop stunt held Dorothy, as well as 20,000 spectators, suspended for two hours and 6 minutes. She flew non-stop, upside down in her open cockpit holding onto the stick with one hand and the wobble pump that kept the fuel flowing to the engine in the other. Those in attendance stood on their feet and gasped as she completed the dangerous series and the *Omaha World-Herald*, reported that "when she landed safely there was an audible sigh of relief from the crowd."[6]

Dorothy's aerobatic records remained unbroken for decades, bestowing a lifelong reign as an aerobatic ace. For such fame and notoriety, she was inducted into the prestigious Pacemaker Hall of Fame at the Museum of Flight at Seattle's Boeing Field. Believing that the 58-year loop record had been hers long enough,

Dorothy suggested that the experienced air show stunt pilot, Joann Osterud, attempt to break the record.

Joann Osterud - Aerobatic Pilot

Joann, a 20-year veteran pilot and air show performer with more than 12,000 total flight hours in her logbook, was prepared for the challenge. Joann performed her first air show back in 1969 in a borrowed airplane. "In those days air shows weren't a big thing. A hamburger, a tank a gas and maybe $25.00 dollars would be our reward for showing up."

Joann remembers "taking off" from the family's kitchen table pretending to be an airplane at a very young age. "I crash landed of course and still have the scar to prove it." She was encouraged by her parents to get her education first, so she waited until she was in graduate school before she started her flight training. "Back then there were few job opportunities for women in aviation. You could only hope to get a good job that would pay enough so you could afford to fly."

In retrospect, it is hard to believe that someone who hated carnival rides and roller coasters almost as much as she dreaded steep turns and stalls, would become an aerobatic pilot. Joann wasn't hooked until a flight instructor with a brand new aerobatic airplane took her for a ride and showed her how to do some basic maneuvers.

Joann's extensive and impressive aviation background covers a wide variety of aviation firsts which explains the respect she has earned within the industry. Not only was she Alaska Airlines first female airline pilot when they hired her in 1975, but she was the first female to compete in the unlimited division at the Reno Air races in 1987. She is also the only woman currently qualified to fly two vintage combat aircraft: the P-51 Mustang and the Hawker Sea Fury. With airline transport pilot, flight engineer and helicopter ratings, she currently flies for United Airlines and still manages to fly a full air show schedule each season. See Fig. 5-5.

Her daring air show routines have thrilled audiences for years and fans mob her airplane to meet the star and get autographs. "As a female stunt pilot I know that I'm somewhat of a curiosity and I think it's wonderful when kids and their parents want to come to see the exciting things a little lady like me can do with an airplane." People are amazed to see the petite pilot crawl out of the high performance airplane, but she quickly reminds all the children that she doesn't do anything that they couldn't do with a little determination and hard work.

In preparation to break Dorothy Stenzel's loop record, certain modifications were made to Joann's biplane. An electric transfer pump was installed to keep the fuel flowing while inverted and special straps and restraints were designed to keep Joann's feet on the rudder pedals and her head from flopping around. She even had a special contraption built so she could hang upside down to build up a tolerance for the prolonged inverted state her body would have to endure.

Fig. 5-5 Joann Osterud.

The outside loop is considered to be one of the more difficult maneuvers because it pulls up to 4 negative G's and forces the pilot's body through the canopy rather than down into the seat. Just one loop is described as a "bone-crushing, gut-wrenching experience" and Joann would have to complete 70 to break Dorothy's 58-year record.

The record-breaking flight was scheduled for Thursday, July 13, 1989 as a prelude to North Bend, Oregon's weekend air show. Low morning clouds threatened the attempt but the ceilings rose to 2,500 feet by mid-afternoon and Joann was set to go. The rules required that the loops be made at an average of one per minute, leaving only occasional seconds for Joann to rollout and readjust the plane's position from the 15 mph winds that pushed her off the judging site.

The weather worsened and the ceiling dropped, forcing Joann to compress her loops to just 800 feet, half their normal diameter. Despite the mounting distractions, after 2 hours, 4 minutes and 38 seconds, she not only broke Dorothy's record of 69 consecutive loops, but shattered the men's record of 180 and did not stop until she had looped a total of 208 times! She had wanted to stop to make it an even 200 but one of her favorite songs started playing on her cockpit tape player and she kept up the beat for another eight loops until the song was over.

As grand marshall for the weekend festivities, 78-year-old Dorothy Stenzel was on hand to witness the attempt and to cheer the new record holder to victory. "It was a double thrill for me to know that Dorothy was there," recalls Joann. "I've always admired her and having her there rooting for me made it easier to break her record." Fittingly, the baton had been passed from one aerobatic giant to another.

For Joann the record setting trend would continue as she zeroed in on breaking a second 58-year record, this one for the distance and endurance of inverted flight. Though Joann's first attempt was foiled in 1990 when the oil drained from her plane's engine, careful planning and additional precautionary measures would spell success in 1991. On July 24, 1991 Joann took off from Vancouver, B.C. in her Ultimate 10-300 S biplane, flew up to altitude and turned upside down. She flew inverted for 4 hours, 38 minutes and 10 seconds, covering 658 miles before she turned right side up to land at Vanderhoof, B.C., setting the new world record.

As Joann's world records rest on the history books to beckon a new challenger, the world can view her Stephens Arko that hangs inverted from Seattle's Museum of Flight at Boeing Field. Joann has laid a path of aviation excellence and aerobatic showmanship for all those who follow in her wake.

Julie Clark - Aerobatic Pilot

In the air show industry, Julie Clark has been a favorite performer of spectators and sponsors since 1981. Part of the thrill of watching any air show is being entertained with the variety and individual style that each performer brings to the sky. Julie flies an inspiring show with an American theme and a patriotic message that will tug at your heart and bring a tear to your eye.

From the "Air Force One" paint scheme on her T-34, to her beautifully choreographed "Serenade in Smoke" which paints red white and blue smoke streams across the sky, to Lee Greenwood's hit "God Bless the USA," it's easy to see why fans jump to their feet as Julie taxis down the flight line waving our American flag. Julie hears the music in her cockpit at the same time the people hear it below and admits that even she gets choked up flying the routine.

Julie started flying in 1967 and completed her private pilots license in 1968. "I used my college book money to pay for my initial flying lessons and worked odd jobs to pay for my advanced ratings," Julie explains. "My first career goal was to become an airline pilot. My dad was an F-27 captain for a regional airline called Pacific Airlines and I used to fly with him on his trips. I idolized my dad, he was my hero and sparked my interest in aviation." See Fig. 5-6.

Julie's father, Ernie Clark, died in 1964 on a flight between Stockton and San Francisco, when Julie was only fifteen years old. The airplane's flight data recorder revealed that a deranged passenger entered the cockpit and shot Captain Clark, causing the airplane to crash, killing all 44 people aboard. "For

Fig. 5-6 Julie Clark

the first time in the history of aviation, my dad, a pilot at the controls of an airliner, was shot by a passenger! Back then the word hijacker didn't even exist, there was no such thing." This accident would bring about the Clark Act of 1967 which has since mandated that all cockpit doors be locked in flight.

Julie reached her airline goal when Hughes Airwest hired her in 1977. Hughes purchased the airline her father flew for and consolidated. Julie was especially proud to be flying for them. "I wore my dad's airline wings inside my jacket my first year and flew with captains who had been co-pilots with my dad." Hughes Airwest merged with Republic Airlines and, they, in turn merged with Northwest Airlines, where she flies today as a DC-9 captain based in Minneapolis, Minnesota. Julie has logged over 16,000 accident-free hours in the air and has flown more than 65 different types of aircraft. Being a success at the airlines and a sensation in the air show industry requires some creative flight scheduling, but Julie easily manages both roles.

As a T-34 instructor at Lemoore Naval Air Station, Julie first fell in love with the airplane she flies today. "Although the T-34 was a military trainer and is stressed for aerobatics, it's not a high performance aircraft and was never designed for aerobatic display," she said. Flying a vintage warbird that is not as maneuverable as contemporary aerobatic planes requires a lot of concentration and anticipation. With no inverted oil or fuel system, Julie must carefully calculate her inverted flight routines. "I do the whole routine with both hands on the stick at all times and often pull as many as 5 and 6 G's."

Julie bought her Beechcraft T-34 "sight unseen" for $18,000 at a government surplus auction in 1977. "I flew the airplane back from the auction site in Anchorage, Alaska to my home in California. Once I got it home, it sat in the weeds because I was so broke I couldn't afford to put gas in it! After I got my airline job in 1977 and a divorce in 1978, the airplane became an obsession and I spent five years restoring it." Julie's painstaking efforts to completely restore the airplane inside and out has brought the market value of her aircraft up to $300,000. Even the original 225 horsepower engine has been replaced by a 285 horsepower, 24 karat gold limited edition Victor engine, which provides better performance.

"It was never my intention to do air shows, never in my wildest dreams." Julie started flying as a part of a three-person T-34 formation team in Harlingen, Texas in 1979. "The two guys I flew with lived distances apart and we didn't practice as much as I wanted so I eventually left the team. I was asked to do a solo routine in Livermore, CA in 1980 and the next thing I knew it was turning into this big business."

Being in the business for the last twelve years, Julie and her T-34 have appeared in some of the biggest air shows in North America, Canada, Nova Scotia and Bermuda. With this exposure and experience, Julie has definite ideas about what makes a good air show. "Variety, that's the name of the game. Team acts, solos by both male and female performers, comedy, beauty, military high-tech as well as warbird appearances are all a must as well as having a top-notch narrator." As much as Julie likes to be invited back year-after-year to an air show site, she believes it's important to rotate her acts so that the performance doesn't become stale.[8]

Julie stresses the importance of flying a "safe" show. "Safety cannot be compromised and should be the performer's number one priority. You don't have to create 'death defying' acts to please the crowds, they're just not necessary. The act's uniqueness and entertainment value is more important." Julie is as enthusiastic as the program she flies and enjoys getting involved with her audiences by visiting and signing autographs after the show. "There is no doubt that air shows get young people interested in flying," she says. Julie strives to inspire one person per air show to get involved in aviation. "If I can do that then I've reached one of my main goals as an air show performer."

Julie is a charter member of the International Society of Women Airline Pilots, a member of the Ninety-Nines, and she is active in the Confederate Air

Force. Julie was also the youngest recipient of the Ninety-Nines' "Woman Pilot of the Year Award" from their Southwest Section in 1980 and was selected to "Who's Who In California" in 1982. She went on to earn "Outstanding Young Woman of America" in 1983. *General Aviation News* selected Julie as their 1988 "Performer of the Year" and their "Favorite Female Performer" in 1988 and 1990. In 1991, her fellow air show pilots presented Julie with the "Bill Barber Award for Showmanship." The FAA recognized Julie with several awards: the prestigious "Certificate of Appreciation" for Outstanding Contribution of Professional Women in Aviation, Contribution of Preservation of Military Aircraft, and Contribution to Women Pioneers in Aviation. In the last decade the latter award has only been presented six times.

Julie's T-34 has also won its share of awards including; "Most Improved Warbird" at Oshkosh in 1981, and the "Ladies Warbird Choice" award at Oshkosh in 1984. Together they have appeared in many nationally televised PBS documentaries and aviation related TV specials like; "P.M. Magazine", "Lifestyles of the Rich and Famous", "ABC's Wide World of Flying" and the game show" To Tell the Truth." Julie has represented Mopar Chrysler Motor Parts since 1988 and flies today as their spokeswoman.

The positive image Julie projects as a professional pilot and performer is a dedication to her love of flying and her pride in being an American.

Sharon Fitzgerald - Aerobatic Pilot

At the age of five she was the second youngest person to climb Mt. Lassen, an extinct volcano. Sharon led her parents, Ed and Marilyn Fitzgerald, up the mountain with a dogged determination. But her parents gently guided her then, as they still do today. On the climb up the mountain they picked spots to stop and pause and look at the climb still ahead. They asked her if she wanted to continue. She did. By nine years of age, Sharon was a skilled show dog handler and by the age of eleven, she had saved a drowning victim. Between nine and eleven, she also decided she wanted to be a pilot. Sharon Fitzgerald is an honor student and took her first flight lesson on her eleventh birthday. On her thirteenth birthday she took her first helicopter lesson. Today she continues to build her flight time in a French Canadian two-place *Robin Sport*. Sharon is also looking forward to the day when she can solo the airplane. That day is about three years away. You see, Sharon is only thirteen years old.

It is easy to forget Sharon's age once you start talking about airplanes. She has met dozens of aviation personalities like Chuck Yeager, Jimmy Doolittle, Jeana Yeager, Patty Wagstaff and astronaut shuttle pilot Major Eileen Collins. She holds her own easily when the conversation centers on snap rolls, Immelmans and lomcevaks.

She met famed former Lockheed test pilot Tony LeVier, who started her on his Safe Action in Flight Emergency (SAFE) program, which is designed to teach students to fly like pilots were taught in World War II.

"She's a better than average student, and is highly motivated," says Peggy Hart, co-owner of Hart Air. How motivated is Sharon? She explains her future this way: "I would like to fly and break records and graduate from the Air Force Academy in 2001. Someday I would also like to be a general in the Air Force but my immediate goal is to climb the ranks of the Civil Air Patrol. And I haven't quite decided whether I want to be a fighter pilot or an aerobatic champion." See Fig. 5-7, 5-8.

Sharon is well-grounded in aviation but that's not all she can do. She intends to be active in charities and in programs to help animals, people and the environment. She is also a strong proponent in getting kids and females into flying. "I would like to see more people getting active in their dreams," she said, "and fulfilling their wishes." Sharon is well on the way to serving as a role model for others in this respect.

Today Marilyn and Ed Fitzgerald still encourage Sharon to explore her possibilities. Like climbing the mountain, the trio occasionally pauses to re-focus on the goal, discuss it and then they usually continue their journey.

Katrina Mumaw - Air Combat Ace

The two Marchetti SF-260s glistened in the bright morning sun. It was clear from the maneuvering that they were engaged in a dogfight. The planes wheeled and rolled and looped through the sky during the fight. Katrina Mumaw was in the lead plane on the defensive. She had successfully evaded the aggressor's gunsight several times, slipping out of the enemy's rectile an instant before he depressed the trigger. After five minutes of violent high-G maneuvers designed to make the other pilot work extremely hard, thus wearing him out, Katrina decided to terminate the aggressor. She banked hard left, and deployed ten degrees of flaps. That chopped her airspeed and she watched as the opposing Marchetti zoomed past. Katrina quickly rolled horizontal and was hot on the tail of the aggressor.

Her opponent was good, he was a Marine Corps F-18 driver, and Katrina was having trouble tracking him. He was jinking and dodging her and kept slipping from her gunsight. The Marchetti was built for a pilot 5 foot 8 inches or taller, Katrina was smaller and this did not work to her advantage. She had trouble seeing everything over the instrument panel sometimes. But this had happened before. She quickly rolled inverted. Now her view was clear and in a moment she fixed her gunsight rectile on the opponent's wing root. She depressed the trigger and a moment later a plume of smoke began trailing from the plane. Another kill, Katrina's 69th. Her previous "victim" was a 1,000 hour "Tom Cat" pilot/instructor, and veteran of the Gulf War. He said after the duel, "My little friend here is no little girl in an airplane." See Fig. 5-9, 5-10.

When asked by a local TV newscaster what she thought of the idea that "combat flying is a man's world only," Katrina quickly replied, "I think the idea is stupid." He then asked her if she thought girls could do it. She smiled and said, "Yes, probably even better."

Fig. 5-7 Sharon Fitzgerald with her instructor Richard Ardenvik.

Fig. 5-8 Sharon Fitzgerald after her first helicopter lesson. Mom Marilyn is on the left. Instructors Rob Carone & Tony (l) of CSR Aviation, in Long Beach.

Fig. 5-9 Katrina Mumaw trying out a museum piece.

Fig. 5-10 Katerina Mumaw seen here in the Air Combat Marchetti SF-260, the plane that made her an Ace.

The possibilities of an aviation career blossomed for Katrina the day she met Jeana Yeager and Dick Rutan, just before the two made their successful round-the-world, non-stop, non-refueled trip. She decided that day that their jobs as test pilots were neat and someday she would be a test pilot, but her ultimate goal would be to reach for the stars and be an astronaut. That was in 1986, when Katrina was three years old.

Six years later Katrina Mumaw had amassed a staggering amount of aviation experience. In 1988, her father took Katrina on her first airplane ride in a 1927 Travel Air. The pilot did a wing over and thrilled five-year-old Katrina. The pilot throttled back and Katrina could hear the wind whistling through the wires of the ancient but venerable biplane. Back on the ground, Katrina's father Jim discovered his daughter was slightly disappointed. Katrina asked the pilot why he had not flown inverted. He replied that Katrina was not wearing a parachute and might fall out.

Katrina's reply was, "That wouldn't matter, Daddy would catch me."

Due to a divorce, Katrina, her little brother, Nick, and their father found themselves on their own. To encourage Katrina, her father started collecting aviation memorabilia that would also serve as an investment for his children. As the collection grew, it became apparent that they would have to be practical. The dining room set and china cabinet found their way to the garage as room was needed for control panels, ejection seats and the like. When Katrina's father came home with the tire from an XB-70 he wondered where this treasure would go. To his amazement Katrina insisted on relocating her Barbie dollhouse into the garage. No matter what he said, her father could not change her mind. So now the huge tire sits in a place of honor in their home.

The most endearing and precious part of the collection is growing geometrically. Katrina's vast collection of photographs sent to her as encouragement from some of aviation's best grows daily. Most are signed and many come with notes and letters. The list is long and Katrina has heard from test pilots, air race pilots, air show pilots, astronauts, pioneer pilots, record-breakers, fighter pilots and a B1-B bomber squadron.

The collection of memorabilia is also growing. Katrina has a J-34 engine in the back yard. She has patches encased in the ceiling since the wall space is filled with photos. There are mannequins in flight suits and even a Russian Mig 25 suit and helmet. Museum directors now come to Katrina's house to examine the family's growing collection.

In 1990, Katrina had her first balloon ride, but when she turned eight, things went into high gear. A trip off Catalina Island afforded Katrina a chance to parasail. She took a helicopter ride over the Christo Umbrellas in Gorman, California. Next she took a glider ride over the Tehachapi Mountains and experienced 4G turns that squashed her into the seat, and shallow turns that lifted her and had her floating in her harness. Next it was a ride in a modified P-51 Mustang out of the Planes of Fame in Chino, California. There she finally got a chance to fly inverted in a series of rolls. She also got to experience a dogfight when another Mustang jumped them during the flight. The stage was set for Katrina to enter Air Combat, but not before her father took her to a hill just beyond Edwards Air Force Base where a photo opportunity of the B1-B bomber was to take place. Imagine the huge bomber bearing down at 680 mph with its wings swept back; the roar as it passes a mere hundred feet overhead, blotting out the sun for a split second and afterburners ringed by flame as the big plane climbs skyward. The wind whistled through the sagebrush eerily as it followed the bomber, picking up dust and small stones with it. Picture this not once, but many times, including low-speed passes as the bomber swings its wings out and looks as wide as a football field. By now there is no doubt where this young lady is going when she grows up..

Katrina is a life member of the Edwards Air Force Base Flight Test Historical Foundation. The group is in the process of getting a museum built to house the growing collection of flight test aircraft and memorabilia. The museum will have as its central exhibit SR-71 "955." A plaque for the exhibit will carry Katrina's name on it - it will be her exhibit.

Katrina is also a life member of the Friends of the American Fighter Aces Foundation and a member of the Experimental Aircraft Association.

Katrina lives about five minutes from the Society of Experimental Test Pilots and the staff is like family to her. Among the photos in the director's office is one of a young lady standing in front of a B1-B, signed by Katrina to the society as a future member.

For hobbies Katrina enjoys reading , both still and video photography (she's won several honorable mentions in photo contests) and gymnastics. She doesn't consider flying a hobby or does she look on it as a job. For Katrina flying is a fact of life.

Katrina pulled an "A" on her third grade report card and eleven "A + "s. She is in the accelerated classes and plans on entering the Air Force Academy after high school in the year 2001. She then plans on flying fighters and going to Red Flag long enough to log the hours needed to get into test pilot school. Her long range goal is to become an astronaut and either be the mission commander or pilot on the first mission to Mars. It is clear that it is more than a coincidence that Katrina will be starting her aviation career and space odyssey in the year 2001.

Cheryl Stearns - Parachutist

Fifty years after "Tiny" Broadwick retired from parachute jumping you still don't find lots of people who have a burning desire to jump out of a perfectly sound airplane. There are several thousand devoted skydivers, however, there are far fewer pilots willing to leave an airplane this way. Cheryl Stearns jumps for joy at the very thought of leaping! Cheryl, who is both an airline pilot and a world parachute champion, spends most of her time either flying in or jumping out of airplanes. This may seem like a contradiction, but Cheryl keeps her priorities in order. See Fig. 5-11.

Fig. 5-11 Cheryl Stearns

Cheryl was 17 when she made her first jump. "I wanted to experience the sensation of falling through the air so I decided to take a lesson. I talked my mother into signing the release form because I wasn't old enough to jump without it. I even borrowed the initial $40 fee from her." Four months later, Cheryl's father got her started on her flight training and that started the unique balance that Cheryl has found in flying and parachuting. She completed all her ratings and started flight instructing while she attended community college in Scottsdale, Arizona.

In 1974, Cheryl entered her first national skydiving competition and placed ninth overall. She knew she wanted to become a champion, so after graduation she moved to Reaford, North Carolina to train with world-renowned jump trainer, Gene Paul Thacker. "I arrived on Gene's doorstep with $50 in my pocket and my dog Chalacko in tow ready to train."

"What really impressed me about her," said Thacker, "what made me encourage her was that at 19 she already had a couple of years of college, many job experiences and had gotten a commercial rating. Anyone 19 years old with those credentials has to have a lot of initiative. Cheryl just has a burning desire to be the best at whatever she does."[9]

Cheryl worked and trained hard by jumping more than six times a day, seven days a week and piloted the planes for the other skydivers. Cheryl knew there was more to training than just learning how to jump, it's style and accuracy that turns a jumper into a champion. "In accuracy training you have to pay special attention to the wind direction that varies from altitude to attitude," she said "you must have precise handling of the canopy to land on a target the size of a silver dollar."[10]

In the style category one must perform a series of aerial maneuvers within a 30-second time span before the chute must be opened. Here an athletic physique is as important as the timing itself. "When you jump out of a plane you don't just drop; an 80-mph wind meets you from the airplane's forward throw and you feel the speed build up around you. You have to work with that speed to perform the maneuvers by digging your hands in the work against it. Balance is also extremely important because without it you could wind up in a flat spin and tumble out of control."

With Thacker's coaching and Cheryl's determination, the training paid off. At the 1975 national competitions Cheryl placed first in accuracy, 11th in style and won the woman's overall title. Cheryl had established herself as a major competitor and was well on her way to becoming a true champion.

In February of 1977, Cheryl made history be becoming the first woman member of the Golden Knights, the U.S. Army's parachute team. She enlisted in the Army and served two tours of duty over an eight year period while training and jumping with the Golden Knight's elite corps. "All my life I wanted to be a world champion and the Golden Knights made that happen for me. The guys were like brothers," she said, "and really took me under their wings. They supported me and encouraged me and never questioned my jumping ability."

Cheryl left the Army in 1985 with the rank of sergeant and a B.S. in Aviation Administration and a Master's Degree in Aeronautical Science from Embry-Riddle's Fort Bragg Campus. She graduated on the dean's list and achieved both of her degrees Magna Cum Laude.

Today Cheryl has logged more than 8,400 jumps and 8,300 hours of flight time and is still in the Army National Guard. She has won 25 national and international parachuting championships, including 30 records and has won the U.S. skydiving championship 13 times, twice beating all competitors both male and female. Cheryl Stearns is the only person in parachuting history to hold four world parachuting records simultaneously. In 1991, Cheryl was selected by the International Parachuting Committee to receive the Leonardo da Vinci Diploma, the highest honor the international sport of parachuting can present to one of its own.

Part of what makes Cheryl's career so impressive is the length of it. "Most athletes have a tendency to peak," says Guy Jones, who was the executive officer of the Golden Knights when Stearns trained with them. "Cheryl just keeps coming back year after year."

According to Thacker, "She's still the best of the best."[11]

Among Cheryl's records are some truly unbelievable feats. As the world record holder for accuracy, Cheryl jumped from a point of 2,500 feet and was able to land dead center on a 4 inch disk 43 consecutive times in the daylight and 23 consecutive times at night. As the world record holder in the style category, Cheryl completed a series of intricate aerial maneuvers during her free fall in only 6.3 seconds!

In 1987, she tripled the previous women's record of 79 jumps in a 24-hour period by totalling 255 jumps by day's end. The previous men's record was 240 which she also broke that day along with fellow parachutist Russell Fish who jumped simultaneously with Cheryl. Twenty-four hours of back-to-back jumping required the coordination of ground support personnel, parachute packers and three airplanes. This support was provided by sponsors who were curious to see if the old records could be broken.[12] "We jumped from 2,000 feet landed on the target and then ran about 30 yards to 'chute up.' The next plane would be waiting to take us back up to start the drill all over again. We ate and drank on the ride to 2,000 feet and port-a-potties were set up for pit stops."

Suspended to a parachute, Cheryl has done more than just experience the sensation of falling."I've jumped out of helicopters, hot-air balloons, and out of the tailgate of a C-141. I've jumped at night, have fallen through snow, needle sharp raindrops, and into bugs and bats!" She twice parachuted onto the 50 yard line at the Fiesta Bowl, once delivering the coin for the toss and the other time with the football in hand, and has landed at the base of the Statue of Liberty carrying our American flag. Cheryl also toured with a barnstorming entertainment troupe called Air Show America in the late 70's, demonstrating her talent and skill.

As a senior first officer for USAir, Cheryl has a flexible schedule that allows her time off every January and February to train with the Golden Knights in Yuma, Arizona. She continues to compete around the world and vows to keep jumping until they have to carry her off the plane! "I'm very lucky to have found a sport where talent and love come together. After 20 years of jumping the thrill is still there and that's one of the things that keeps me at it. The other is the camaraderie of the sport and the friendships I've made around the world, that's really what keeps me jumping!"

As "Tiny" Broadwick found, there are only so many things a person can do with a parachute. But unlike Tiny, Cheryl becomes innovative and makes it fun. Once she had a friend fly an open cockpit Pitts Special upside down so she could experience falling out of an airplane. "I unhooked my seatbelt and put my feet on the seat," she says. "He did a loop, held the plane upside down and I dropped out. It was funny, I didn't jump out, I fell."[13]

Misty Blues - Skydiving Team

Tiny Broadwick may have been unable to devise new and exciting parachute routines, but she was limited by the technology of the times. Today, there is a group of women who have dozens of exciting jump routines in their repertoire. The Misty Blues have taken Tiny Broadwick's freefall and developed it into an art form.

The sport of skydiving seems terrifying for the beginner and thrill-seeking to the novice and although many people attempt it, few actually master it. The exception to the rule is the well-known all-female skydiving team called The Misty Blues formed by Sandra Williams in 1985. The 14-member team performs nationally and internationally, entertaining their audiences with a parachuting spectacle of aerial stunts and pageantry. See Fig. 5-12.

Although more and more women are taking up sport parachuting today, statistics show that only 10-12% of skydivers are women. With less than 1,000 women skydivers in the United States, the Misty Blues make up a very select group of women. Membership on the team is by invitation only and strict requirements apply.

Candidates must have logged at least 1,000 jumps and be willing to try new and wild stunts. They must be brave and have a confident and congenial personality and there is no tolerance for drugs. "Some of these requirements may sound strange," says team leader Sandra, "but as a team we spend a lot of time together practicing and performing. We want to be able to enjoy each others' company and have a good time doing what we do."

As a result, the Misty Blues are not only one of the most unique and exclusive groups of women, but also the most experienced. The team has more than 30,000 jumps collectively and each member holds at least one world parachuting record. Spread throughout the country, the Misty Blues also maintain full-time professional careers between training, traveling and performing.

Team member Dawn Ford is an elementary school teacher, Kay Greip is a personnel director, and Bev Hein owns her own business. Renee Daniels-Hunter is a corporate branch manager for the "Graingers," Bambi Knight is a registered gemologist and Vee Kramer is a commercial artist. Nancy Kurlin is a cardiac intensive care nurse, Jill Martain is a network switching technician, and Susan Perkins is a vice president and director of research for the largest stock firm in the southeastern United States. Martha Scott is a field engineer, Gail Sims owns her own skydiving school and team leader Sandra Williams is not only an artist but a nationally-rated skydiving judge. Elaine Turba is a comptroller, and Anne Worden works in regulatory affairs in the FDA. One would wonder where these women find time to fit skydiving into their busy schedules, but jumping is a shared passion that they make time to accommodate. See Fig. 5-13.

Every spring the team members get together at their headquarters in Florida to attend an extensive training camp. The ladies work together closely, perfecting the many skills that are required to master the routines. The popularity of their performances comes from the wide variety of exciting stunts designed and

Fig. 5-12 The Misty Blues headed earthward

choreographed by group members themselves. These original maneuvers require special equipment and Sandra's husband, Gencarelle, is the genius who devises the systems, designs the harnesses and manufactures the rigs for each stunt.

Besides carrying a 50 pound, 800 square foot American Flag or a 75 foot long rainbow banner which is unfurled from 3,000 feet at the beginning of each of their performances, they execute stunts called Banner Ballets, Parachute Dogfights, Pinwheels, Side-by-Sides, Down-Planes, Stacks, Trapeze Stunts and Rainbow Smoke Chains all while dangling from their trademark pink and blue canopies. Stunts routinely begin from 5,000 feet and with little time to waste they must go straight into their routines with no free fall.[14]

The most exciting stunt performed by the Misty Blues is the intentional "Cutaway." This maneuver simulates the disaster of being attached to a faulty parachute, the worst malfunction a skydiver can experience. The jumper purposely releases one side of the canopy causing it to spin violently, seemingly out of control. This puts the diver through a ride that would rival any roller coaster. The diver must hold on tight until she can eventually release the malfunctioning chute and go back into a free fall and deploy her main chute. The stunt looks extremely dangerous but is actually a safe maneuver that each member is trained to do. Misty Blues with strong stomachs consider this to be their favorite stunt.

The Misty Blues can be seen performing their exciting and exhilarating shows everywhere from air shows and balloon fiestas to weddings and sporting events. They have traveled around the world on exhibition and have even performed for royalty. "In Indonesia we performed in a large air show mainly as role models for Indonesian women." says Sandra. "They all enjoyed the show but thought the blondes on the team were from another planet! In 1991, we were in

Fig. 5-13. Six members of the Misty Blues: Back row (l), Sandra Williams, Renee Daniels-Hunter, Bambi Knight, Anne Worden, Front (l) Elaine Turba, Dawn Ford.

Malaysia and jumped with the Royal Malaysian Air Force in the first air show that country had ever organized. Jumping out of a Russian Kazan MI-17 helicopter with no seatbelts or doors had to be one of our more memorable events, especially when they wouldn't let our interpreter wear a parachute."

In 1991, the Misty Blues were a part of the largest all-women skydiving event ever recorded in the world. A total of 88 women skydivers from all over the world jumped together and formed a aerial cluster chain that was held for a total of 2.87 seconds. This large international female consortium included skydivers from the U.S., Japan, England, Australia, New Zealand and all parts of Europe.

With all the talent, experience and showmanship of these women, they were honored by performers, convention organizers and sponsors of the 1991 International Council of Air Shows convention by being voted the "Best of the Best" of all civilian skydiving teams in the United States. Being on The Misty Blues team is widely regarded as an impressive addition to any female skydiver's resume and with standards so high, it's easy to see why they truly are the best of the best. See Fig. 5-13.

Jessie Woods - Wing Walker

In 1991, at the age of 82, Jessie agreed to come out of retirement after 53 years, to wing walk again at Lakeland, Florida's Sun'n Fun. All eyes were on Jessie, the grand lady of wing walkers, as she took to the skies on the wing of a 1928 Standard. "I was a little apprehensive at the thought of doing it again after all these years, but once I got out there it felt like I'd never gotten off those wings."

Jessie Martin eloped with a charming, tall and handsome young aviator named Jimmie Woods in her teens. She traveled the countryside with him, buzzing towns, selling rides to the locals and leading the nomadic lifestyle that barnstormers do. Jimmie taught Jessie to fly during those early days. "If the field we were in was safe and if we had a few extra dollars for gas, Jimmie would let me shoot some landings." Turning their talents into a way of eking out a living, Jimmie and Jessie formed their own flying troupe called the Flying Aces Air Circus in 1929.They flew throughout the country with an entourage of pilots they hired to perform air shows and exhibitions. See Fig. 5-14.

The depression era brought lean years for everyone, including barnstormers and pilots. The Woods were forced to come up with creative ways of attracting the crowds. "One night the guys were sitting around trying to come up with a new idea for the show and one of them said, 'What we need is a wing walker, a woman wing walker' and being the only gal in the outfit they all looked at me." None of them knew anything about wing walking or the aerodynamics that were involved, but at nineteen, Jessie agreed to give it a try.

Jessie recalls her first attempt at wing walking to be her most terrifying. *My husband Jimmie rigged a rope that he tied around my waist and attached the other end inside the cockpit. He got up to altitude, throttled back and I stood up. Stepping into the slip stream took my breath away and from that moment on I was scared*

Fig. 5-14 Jessie Woods

to death. I walked out on the wing and tried to slip between the spar and the guide wires but the rope that I was attached to got caught in the wires. I couldn't move in any direction and thought the world had come to an end. Finally I got mad, because I was kind of forced into it doing this in the first place, and anger overcame my fear. I untangled myself and made it back to the cockpit. When we got back on the ground, Jimmie and I had our first fight ever and I won! If I was going to do this, I was going to do it my way and I never again wore any ropes or restraints while wing walking.

The act was billed under Jessie's maiden name, Martin, because it sounded better to have a glamorous, unattached wing walker to thrill the crowds rather than the boss' wife. She wore her hair in a long braided ponytail and wore different costumes, everything from a red bathing suit to an all white jumpsuit. "I had to stop wearing the bathing suit because in order to get out on the wing I had to climb up over the airplane's exhaust stack. The heat would scorch my legs and they were constantly burned and scabbed." See Fig. 5-15.

Jessie was a professional parachutist, but never wore one in her routine unless being photographed or filmed. "To wing walk without a parachute was illegal but without pictures the Civil Aeronautics Authority (CAA) couldn't prove that I didn't have one on," explained Jessie. "Parachutes were bulky and caused a lot of drag. It was hard enough to slip through the guide wires out on the wings much less to squeeze a parachute through, too."

Fig. 5-15 Jessie Woods

One of Jessie's wing walking mishaps came the day that Paramount studios came out to film footage of Jessie's act for their newsreel. "Since I was going to be on film I had to use the parachute. We did the routine over and over to please the camera men. They kept saying, 'do it again, do it again,' and I guess I just got tired. My hands got numb and I lost my grip and suddenly I realized I was loose. I opened the parachute and landed safely in a cow corral, manure and all." Wearing a parachute saved Jessie's life that day.

"The first few years of our show was a period when 'anything' in aviation was spectacular. The towns and people were scattered and isolated. There were few roads, railroads, busses or airports. Imagine the effort for a family to drive 50, 60 or 70 miles in the slow cars of the era, and no filling stations or cafes on the way. And yet we drew the crowds at a dollar a carload and managed to succeed. There were, however, some missed meals in the beginning years."

"The best airplane to wing walk on was a Stearman or a Swallow. I'd walk back and forth on the lower spar always having a firm grip in one hand before letting go with the other. The Travel Air was best for aerobatic work." Jessie would sit on the top wing over the center section while the pilots did spins, loops and rolls. One of Jessie's routines included a rope and ladder act where Jessie would climb down from the landing gear to dangle like a trapeze artist from the bottom rung.

"During one of those rope capers my blouse got caught in the wind and actually ripped off of me. To make matters worse, my brassiere strap snapped and flew off, too. When the act was over and I started to climb back into the cockpit, the pilot saw only skin emerging from below. He realized my predicament and ducked his head inside the back cockpit until I had climbed in the front.He even took his own shirt off and handed it up to me. I thought I'd gotten away with the whole ordeal until one of the spectators came up to me after the show shaking a telescope in his hand telling me how much he'd enjoyed my act!"

Jessie retired from wing walking in 1938 when the Flying Aces Air Circus was disbanded by the CAA. The longest running and most successful air circus after nine consecutive years was forced to close because the CAA was entrenched in promoting air transportation. They suppressed entertainment troupes like Jimmie and Jessie's, believing they were not a "safe" form of aviation.

From a wing and a prayer to a wing and a dare, legendary wing walker Jessie Woods spans the generations with the spirit that immortalized the glorious aviation era of the 1920's and 30's.

Cheryl Rae Littlefield - Wing Walker

In 1980 Cheryl Radtke watched her TV and saw a newscaster walk on the wing of Gene Littlefield's Stearman to promote a local air show and thought, "If he can do, I can do it!" A mutual friend heard her comment and called Cheryl to announce that Gene was looking for a new wing walker for his show. "It was the opportunity of a lifetime and I had to put my money where my mouth was. This was my chance and although I wasn't sure I wanted to make a career out of it, I knew I had to give it a try.

Cheryl called Gene to express her interest and found him at wits-end having been inundated with calls from those who wanted the job. "I'm sure he thought, 'oh great, another one who thinks she wants to try this.'" Cheryl got up her nerve and drove out to the airport but missed meeting Gene. He wasn't scheduled back for two hours and instead of leaving, she found she just couldn't seem to tear herself away. "When Gene returned we talked about the whole procedure and it made me feel less apprehensive, that is until he took me into the hangar to see his airplane. He was in the process of re-covering the wings and the plane was in a million pieces all over the floor. I thought, what in the world am I getting myself into."

In the months that it took to re-cover the Stearman, Cheryl hung around the hangar, learned the ropes and got comfortable with the environment. Gene used the time to teach Cheryl everything she needed to know and prepared her for her first attempt at wing walking. "Gene had already carried one female on his wings and knew what to expect. He relayed previous experiences to me and explained what everything would feel like. By that point I had overcome my fears and anxiety. I had great confidence in Gene's ability and was ready for my first ride."

Gene took off with Cheryl on the wings and flew around the pattern a couple times before bringing her in for a landing. "I was disappointed and thought if this is all there is to it this is pretty boring!" "Well I haven't turned the airplane upside down yet," replied Gene, and with that she insisted they go back up and they performed the entire aerobatic routine with Cheryl aboard.

Cheryl's parents had to be tricked into seeing her wing walk for the first time but even Cheryl's mom was ready to give it a try by the end of the day. "After a few months of practice I made my air show debut on July 4, 1981 at Barnett Field, in Mt. Morris, Illinois. My folks were there to watch but weren't familiar with Bob Heuer's comedy act where he works the crowd pretending to be a drunk. Before we were scheduled to go up, Bob staggered out to our airplane and started shaking the pylons. My dad was furious that this drunk was fooling around on wings his daughter was about to walk on and started out across the field after him. I had to stop him and explain that it was part of the show." See Fig. 5-16.

Cheryl and Gene spent many hours practicing and perfecting their routine and formed an unusual sense of trust that bonded them together. Their relationship grew from their professional lives into their personal lives and they were married in 1986.

Fig. 5-16. Gene & Cheryl Rae Littlefield

Today Cheryl is one of only ten wing walkers in the world that actively performs the air show circuit. Climbing the wings of their modified 1942 Stearman dubbed "Magic One," Cheryl has a personalized style and a specialized routine that features many stunts and a crowd-pleasing disappearing act. "Gene makes a low pass over the crowd with me visibly on the wing then he turns and flies back doing a four point roll overhead to reveal that I have mysteriously disappeared. On the final pass I am positioned safely back on the wing and we wave good-bye. People stop by to greet us after the show and want to inspect the airplane to see how the trick is done. Usually the kids are the ones that figure out my secret."

Although Cheryl uses a specially designed safety support system when she is on the aircraft's top wing, she is not attached when moving about the airplane in flight. It takes great dexterity and a unique athletic ability to ride through the maneuvers and safety is their biggest concern. Gene has served as the president of the Professional Air Show Performers Association and the International Aerobatics Club. He has also been an aerobatic competency evaluator for the FAA and has had a hand in establishing a resolution which prohibits air show members from directing energy within close proximity of crowds.[15]

Cheryl recalls one mishap when the engine quit at low level while inverted with her perched on the top wing. "I actually felt my feet leave the airplane. Fortunately Gene found enough airspeed to turn the plane right side up and we landed in the grass between the runway and the taxiway." The incident hasn't quelled Cheryl's wing walking ambition. As a fixed base operations office manager for A & M Aviation at the Clow International Airport in Plainfield, Illinois, Cheryl has a flexible schedule that allows her to travel and perform at least 20-25 air shows per season. She is also a licensed pilot and enjoys the ride almost as much from behind the controls of the airplane looking out over the wings, as she does walking on top of them.

Lori Lynn Ross - Stunt Woman/Wing walker

Lori Lynn Ross is a professional Hollywood stuntwoman who was recruited by veteran aerobatics pilot Jim Franklin to be an airplane wing walker. The petite 4 foot 11 1/2 inch, 98 pound daredevil has been involved in action and adventure sports since childhood when she began competitive springboard diving in high school. "I had to be on the boys' diving team because there wasn't a girls team. Back then I was terrified of heights and the diving coach had to push me off the three meter board because I wouldn't dive off!" Diving became Lori's main sport in college and after graduating with a degree in education, she became a professional diver with the U.S. High Diving Team. She traveled to amusement parks throughout the West and performed live stunts that included clown dives and fiery leaps from thirty feet. Lori met her future husband Dan Hershman in 1981, and he sparked her interest in mountain climbing. It was her diving and climbing ability that led industry friends to encourage her to pursue a career in stuntwork.

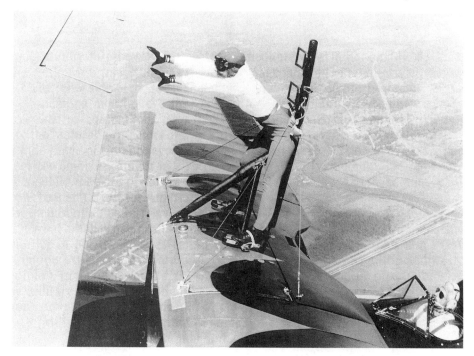

Fig. 5-17 Lori Lynn Ross seen here without restraints

As a member of the United Stuntwomen's Association Lori, has performed over a hundred stunts for commercials, television and motion pictures. When she is not shooting films, Lori brings her unique talent and experience to the world of airplane wing walking. Those fortunate enough to catch one of her acts watch a master in action. See Fig. 5-17.

Lori hangs from the wires between the wings for take off then climbs to the top wing rack for the aerobatics. Her tiny muscular frame causes minimal drag for the pilots as they take her through a full routine of spins, tail slides, four point rolls and Cuban eights. Hollywood stunt work would be considered tame compared to the rigorous workout Lori endures on the Waco's wings, pulling up to 4 G's on some maneuvers at speeds in excess of 200 mph. An inverted pass above the crowd with Lori dangling from her feet is proof of her fearless confidence.

For some shows Lori and her pilot, Eliot Cross, team up with Johnny Kazian and Jim Franklin for one of the most original and spectacular wing walking stunts every performed. The duel wing walking act is highlighted by a breathtaking baton pass between wing walkers as the planes are suspended in mirror flight.

Lori believes that wing walking serves as a unique contrast to her stunt work. "On film if a stunt doesn't come out right we can shoot it again but when I'm wing walking in front of a live audience I only get one chance to do the routine right. My wing walking experiences have sharpened my instincts and made me

Fig. 5-18 Lori Lynn Ross

very aware of the importance of proper equipment rigging. It can mean the difference between a stunt being safe and exciting or disastrous. I take a lot of pride in having a safety-first attitude and a cool head.

"People used to think I was crazy for not wearing a harness or restraining device while wing walking, that is until a failed brake on landing caused the airplane to ground loop. Unattached I was able to jump clear of the accident to keep from being pinned beneath the airplane and most importantly I was able to stay out of the propellers way." See Fig. 5-18.

Currently Lori is developing another stunt involving a plane to plane transfer. She completed her skydiving certification in preparation for the attempt and has already progressed to the point of making physical contact with the exchange plane. Only an accomplished stuntwoman would attempt such a feat. Once perfected, the pinnacle performance will be the supreme culmination of Lori's stuntwork skills and her wing walking talents.

Pat Wagner - Wing walker

Probably the most recognized female wing walker today is Pat Wagner who is only half of the Wagner team that has been thrilling audiences on the air show circuit for more than twenty years.

Pat was the only girl taking flying lesson at the South Dayton Airport Flight School when Bob was the manager in 1966. They met, fell in love and were married in 1968, taking off for their honeymoon in a plane with a "Just Married" sign and tin cans tied to the side.

Their barnstorming and wing walking act started quite by accident in 1971 when the Wagners agreed to fulfill an air show contract for a student who was not experienced enough to fly it himself. When the wing walker they had lined up failed to show, the Wagners figured the "show must go on" and Pat climbed on top of the airplane and prepared for the ride of her life.

"I was scared to death the first time I did it. Everything was going through my mind. It was a learning process for both of us. Having me on top of the wings caused the airflow to change drastically, especially during a stall." The impromptu stunt was such a success that they incorporated the wing walking act as the grand finale to Bob's own aerobatics routine.

Fig. 5-19 Pat Wagner atop a wing

With Pat riding high above the 1940 red and white Stearman, Bob flies her through eight minutes of heart-stopping aerobatics, including loops, rolls, hammerhead turns, Cuban eights and inverted flight. They maintain a busy schedule of 20-25 air shows per season and confess that they don't perform to live but live to perform. They truly love what they do and it shows. Their air show bookings have taken them from coast to coast in the United States and Canada as well as an exciting tour in Venezuela, a highlight of their career. "The people in that country were simply not accustomed to seeing an airplane or a performance like ours. We were known as "Los Aeros Locos." See Fig. 5-19.

Both Pat and her husband have their commercial pilots licenses. Pat has a single engine land and sea rating as well as her glider rating. She has logged more than 1,000 hours, 100 of which are in Waco biplanes. In 1982, the Wagners purchased their own airport, now appropriately named The Wagner International Airport. The acreage includes a 1,900 foot grass strip, a log home and two large hangars that house the Wagner fleet of airplanes. The roof of the hangar is a red and white checkerboard design that matches their Stearman motif and is an easy landmark for cross-country fliers to spot.

Together they can be seen practicing new routines right in their own backyard as the Wagners make some revision to their show each season. After bringing thrill and excitement to our air show skies for more than twenty years, Pat aptly earns the title "Queen of the Wing walkers."

Ballooning

The first woman to go aloft did so in a balloon more than 200 years ago and women have remained actively involved in ballooning. A balloon even carried the first woman into the stratosphere. On October 23, 1934, Jeannette Ridlon Piccard, and her husband, sailed to the height of 57,579 feet over Lake Erie. The flight, however, was not without start-up problems.

Jeannette and her husband Jean were scientists who studied atmospheric conditions and were not pilots nor did they have balloon licenses. The couple decided that Jeannette would get the license and Jean would be the scientific observer on the flight. Jeannette took her first flight on May 15, 1934 and soloed a month later. When she was asked later why she chose to get her license rather than hire someone, she replied, "How much loyalty can you count on from someone you hire?" She was also asked why she did not take parachute training, "If on the first time you jump, you don't succeed, there's no use trying again."[16]

The next problem they faced was financial backing. The 600,000 cubic foot balloon, the one used during the 1933 World's Fair to set an altitude record, was donated to the Piccards by the fair promoters. The traditional backers would have no part of a venture that had a woman as the pilot. Jeannette Piccard said later, "The National Geographic would have nothing to do with sending a woman - a mother - in a balloon into danger."[17] Private backers eventually came up with the necessary funds.

At dawn's first light, the balloon lifted off on its historic flight. Jean Piccard and a pet turtle rode inside the gondola, taking scientific measurements and Jeannette stood outside, on top of the Gondola. The balloon rose into the stratosphere and eight hours later settled safely back to earth.

Jeannette had flown higher than any other woman and her record remained untouchable until Valentina Tereshkova went into space in 1963.[18]

Fear was an issue for some reporters and entered into the questions asked of Piccard. Her reply was eloquent. "Even if one were afraid to die there is so much of interest in a stratosphere trip that one does not have time to be afraid. It is too absorbing, too interesting. . . When one is a mother, though one does not risk life for a mere whim. One must back up emotions by cold reason. One must have a cause worthy of the danger. In times of war one sacrifices self and children for country. In times of peace the sacrifice is made for humanity. If one can forward by so much as even a little bit the sum total of man's knowledge one will not have lived in vain.[19]

Jeannette Piccard also became the first woman to be ordained a minister in the Episcopal church.[20]

Tucker Comstock - Cameron Balloons

There is an ancient dream among humans, the dream of flight. To lose the bonds of gravity and float aloft, delighting in the dance of weightlessness - for ages, countless dreamers have dreamed this dream. Hot air balloonists live it.

Tucker Comstock is one of those who lives this dream. She received her Commercial Lighter-Than-Air Balloon Pilot rating in 1972. There is a small note of irony in the fact that Tucker is a pilot. She had always been uncomfortable in elevators, airplanes and flying in general. Ballooning, however, is different. "I think it happened in the womb," she said. "As soon as I saw a balloon I knew that's what I wanted to do. Ballooning is not like flying in the traditional sense. There are no vibrations, or noise, just an overwhelming sense of peace and serenity. For me," she says, "it's a communion with nature. A balloon is very stable and calm, so riding in one gives me a supreme feeling of peace."

Wanting to get into ballooning slowly, Tucker and her husband, Bruce, started designing and building balloons from the basement of their home. Today Cameron Balloons the name of their company, earns about $3 million a year in sales. And Cameron Balloons is now part of a larger company, Bristol Balloons, in England. Bristol is the largest balloon manufacturer in the world and Cameron Balloons is one of the largest in the United States.

Among one of the unseen benefits and markets for Tucker's hot air balloons, she discovered, was the untapped market in Latin America. Tucker virtually introduced ballooning to Costa Rica single-handedly, and she has trained all their pilots. "It's funny," she says, "but one would think that because of the Latin machismo culture a woman would have a difficult time there. It's been just the opposite, I never had a single gender-related problem. I have been accepted strictly on the basis of my ability." .

Ballooning has been so well-received in Costa Rica that Tucker has just started a venture tourism company called Serendipity Tours.

Tucker Comstock has taken part in ballooning events in Japan, the former Soviet Union, Latin America and Europe.

Tucker also helped the United States in its relations with the former Soviet Union. On a trip to Russia in 1989, she introduced parasailing to cosmonaut/test pilot Magomed Tolboev. Before she left, she donated several hundred yards of fabric to the Soviets so they could make their own parasails. It is coincidental that the word for sail in Russian is "tucker." Today the Russians parasail for both training and pleasure and the parasails are called "Tuckers," in honor of Tucker Comstock.

"I've never flown anywhere that I haven't ended up making some very good, long-term friends," she said.

Those friends in the former Soviet Union resulted in Tolboev and two other Soviet test pilots being invited to attend a meeting of the Society of Experimental Test Pilots here in the United States. A new era in communications and sharing aviation knowledge had been opened because of Tucker Comstock.

Ballooning is not just for politicians, test pilots and commercial use, Tucker says. "It is very much a family sport. "Because it requires a team, you can put kids as young as five or six to work as part of the team, and they become equals with their parents or other team members."

On the serious side, Tucker Comstock has strong feelings about women looking to aviation as career. She says, "go for it" to any woman thinking about an aviation career. She believes competence will always overcome prejudice. Her best example of this is her Latin American experience. "Women can do with competence what men did in the good ol' boy network."

Tucker says she has never seen a competent woman suffer from discrimination. "Incompetent men have been allowed to succeed but they don't stop competent women.

"Ballooning is one of the most rewarding experiences a person can have," she says. "It's like having a house on the top of a mountain, only every day someone comes and changes the valley below you."

Beth Sullivan - H.O.F. Balloons

In the mid-1970s, Beth Sullivan attended a balloon rally. The big, colorful, silent air bags intrigued her. "The colors were so beautiful and it was breathtaking to watch the heat and air make the whole thing come alive. I knew this very first time I saw a balloon that I had to have one," she said. So Sullivan earned her first ride by working with the ground crew three or four times. The first time she went up, she was hooked. This was a great sport, she thought. "I was very lucky when I was first introduced to ballooning. There were a couple of balloonists who were gracious enough to let me join their balloon crews and learn about this great sport. They really didn't know me very well but saw I had an interest and helped me pursue that interest. The ballooning community has

Fig. 5-20 Beth Sullivan

opened many doors including new job opportunities, but most of all new friends." Fig. 5-20.

By 1977, Beth had her private balloon rating and a year later she had her commercial balloon license.

One of her first balloons was a tethered 30-cubic foot one-person balloon. This is about as small as balloons come and instead of a basket it had a seat. It was all white and she called it "Casper."

By 1980, she had become friends with Tucker Comstock and her choice was a Cameron Balloon. She then became a sales representative for Cameron Balloons, and soon began selling rides.

By the mid-1980s Beth was exploring new sky trails in balloon aviation. She had become a certified balloon instructor, and began training students (one of her first students was her husband who today also has a balloon rating). "But we both have different flying styles," she said. Perceptions of Beth also began to change.

"I had a few career difficulties in the beginning," she said. "No worse than the everyday male-dominated world, however. In the 1970s the men looked on my ability to fly a balloon as a novelty. Some said it was refreshing to have a woman in balloon aviation." But actually earning a living in balloon aviation was a different story and put Beth in a different light. "Some men don't think a woman knows enough about aviation as they do. They also perceived an economic threat."

Beth overcame the difficulties, and today Beth Sullivan not only teaches students to fly balloons but she does commercial advertising. Corporations hire her and her balloon, put their names on it and Beth sails across the sky.

In 1990, Beth and her husband, Maury, were one of five teams representing the United States at the First International Hot Air Balloon Meet held in the Soviet Union. Since they own and operate a dealership for Cameron Balloons, Cameron selected them to represent the company on the trip.

In preparation for her trip, Beth and Maury took classes in the Russian language. This involved learning a new alphabet. Later, Beth found body language and facial expressions were sometimes more helpful in communicating.

Once in Russia, after what seemed like an endless trip and countless stops for security checks, Beth realized, "As Americans I don't think we realized how difficult life is there until we experienced it." The major thoroughfares were for the most part poorly maintained. It was impossible to move with any speed. On their bus ride to Leningrad everything moved so slowly that Maury remarked, "The only thing we passed were an ox cart and a three-legged dog."

They were joined by other balloon teams from France, Germany, Belgium, Japan, Lithuania, England and Switzerland. The Russian people responded with warmness and openness. The American souvenirs they brought over, red vinyl frisbees, were an instant success too.

The Sullivans launched their balloons from the picturesque palaces built for the czars. The immaculate facades of the palaces were in sharp contrast to the profoundly neglected courtyards where the balloon landed. The average Russian or tourist never sees the courtyards but from a balloon, new vistas are constantly revealing themselves.

A Russian military truck, driver, civilian school teacher/translator and two crewmembers were assigned to each balloon crew. Sullivan remembers after one of her 20-minute balloon flights it took the ground crew six hours to locate her, despite the two-way radios and maps they had. The Russians, it seems, were not too familiar with their own road systems. The high cost of gasoline and lack of automobiles severely restricts the average Russian and where they travel.

The Sullivans reflected on what was, at the time, a trip of a lifetime as an experience few visitors to the former Soviet Union could duplicate.

While ballooning for Beth started as a hobby and sport, it has mushroomed into an entrepreneurial business. However, a lot of Beth Sullivan's balloon involvement has also been as a volunteer for local and national organizations. She is a member of the Northeast Ohio Balloon Pilots Association, and one of her larger challenges was to chair their safety seminar during 1991 and 1992. In 1986, she and her husband started a local balloon event, the Pro Football Hall of Fame Balloon Classic, in Canton, Ohio. The event is now in its seventh year and draws more than 50 balloons from across the United States. Beth has also served as general chair of the event in 1989 and 1990, and she was Balloon-meister in 1991 and 1992. Beth is also a member of the Balloon Federation's Education Committee.

Before Beth became a balloon pilot she was a paralegal. Would she go back to the earthbound job? Not a chance. "Today if a woman believes in herself and in her possibilities - she should go out and do it. Society will not prohibit you if you have the desire." Beth characterizes her involvement with ballooning this way: "Since ballooning gives me such a delight, it is only fair to give something back to ballooning. I've become involved with the continuing safety and education of balloonists and the organizing of balloon events. I hope I can give back to the ballooning community as much as it has given to me."

Katherine Wadsworth - Kat Balloons

Some people have described Kathy Wadsworth as an intense woman who seems to live by the relentless logic of the totally committed. She does things all the way or not at all - no halfway measures, no compromising. There is no doubt that the ballooning business has its ups and downs from day to day. All flights are subject to weather conditions. It can simply be too windy, too foggy, or just plain too hot to fly. But Kathy Wadsworth also defies geographical odds. Wadsworth's balloon company, Kat Balloons, formed in January 1985, flies the lush green landscapes of the Connecticut and the Farmington River Valleys. The other most notable areas in the balloon industry are in the Southwest. See Fig. 5-21.

Her obsession with ballooning started in 1974. For the first eleven years Kathy and a partner built balloons at home. One of these, a 128,000 cubic foot, 112 foot long dirigible, they called the "Albatross."

As her interest in balloons grew, so did the types of balloons she flew. The team of two built an AX-1, (8,450 cu. ft. 26 feet tall) and an AX-2 (smallest size categories for hot air balloons) which Kathy flew to set 12 world and national hot air balloon records in the AX-1 and AX-2 categories. She set an altitude record of more than 11,000 feet, a distance record and an endurance record. Kathy is the first pilot to make flights in this size hot air balloon. For their achievements, Kathy and her partner were nominated for the Montgolfier Diplome, the highest award of recognition bestowed in the worldwide ballooning community by the Federation Aeronautique International.

In all, they built 35 hot air balloon systems together - the baskets, burners and envelopes, and two hot air airships (steerable balloons). In fact, the twosome built the first experimental hot air airships and were the first to build and promote the small singleplace hot air balloon systems.

During this time the two partners started using balloons for advertising promotion, instruction and charter work. Between 1979, and 1982 their charter work turned into a tour company. Kathy and three entrepreneurial friends operated Haute Voyage, Inc. in the scenic 50-mile long area of the Hudson River Valley. They flew two and sometimes three balloons and had as many as 16

Fig. 5-21 Kathy Wadsworth

people on a tour. The tour lasted from three to five days and based itself in several different locations in the area. Tour guests traded places between the balloon basket and riding "chase" with the crew. This way each person got the full experience of ballooning. The four Haute Voyagers also made sure the groups stayed at some of the most elegant inns and visited many of the historic and famous mansions in the area.

"It takes a big effort to earn a living in aviation, Kathy admits. "I do it only because I have to be in the sky - because of this need or fascination to be in the sky."

In 1981, Kathy Wadsworth's balloon involvement took a new turn, south. One day in early spring a phone call came from Jorge Delano, a businessman in Venezuela. Delano would play an important role in her professional life. He wanted to order a balloon and take lessons.

"He ordered the balloon over the phone and was up here shortly thereafter to begin his lessons. He commuted during the next seven months from Caracas, Venezuela, to Connecticut for balloon lessons.

Once this gentleman had his balloon license he went back to Venezuela. "When he got there he ordered another balloon and invited Wadsworth down to South America to help him put together a trip over the famous Angel Falls. "Not exactly ballooning territory," Kathy pointed out to him.

Delano hired a helicopter, a support team, filmmakers, photographers and an award-winning writer. Kathy launched from the depths of the 5,500 foot jungle canyon to fly over the Falls, the tallest waterfall in the world (3,121 feet).

In 1983, Kathy and the American team went back to Venezuela to fly over the highest peek in the Venezuelan Andes, "Pico Bolivar." Atop "Pico Espejo" at 15,000 feet MSL she floated over "Pico Bolivar" at just under 17,000 MSL. The trip was dangerous. The pyrometer, used to measure the inside temperature of the balloon was in the red soon after the launch. Kathy also noticed that an area in one of the upper panels of the balloon appeared to be changing color; darkening in color and she wondered if the top could be melting.

"The balloon was overheating and we were getting knocked around by downdrafts. Because the balloon was so hot we couldn't blast ourselves to equilibrium," she said.

They felt the balloon heading downward but she couldn't apply heat too abruptly because her pyrometer was so far into the red. As they descended, they pushed east, away from the dangerous peaks, and down a steep slope. The slope was miraculously free of dangerous rocks, and Kathy was able to set down on the only possible landing site for 100 square miles.

The film of this epic flight of two balloons in the Andes, produced by the Lost World Balloon Society, is called "The Edge of Earth and Sky."

Kathy's next balloon trip found her floating over the archipelago known as "Los Rogues," a chain of 135 coral reef islands. This time she launched from the deck of a cargo ship in a 20 mile per hour wind.

Kathy has flown over the mighty Orinoco River, grazed through monkey-filled treetops in the endless LLanos and from the edge of the deepest and largest sink hole in the world, 1,000 feet across and 1,000 feet deep, found near the edge of the Sarisarinama region of the Amazones in south central Venezuela. When Kathy stood at the base of the sink hole she was only one of eight humans to ever set foot there.

From these several unforgettable balloon flights over some of the most challenging landscapes in South America came two films: "Above the Lost World" and the latest from many visits between 1985 and 1988, "At the Edge."

"Above the Lost World" has enjoyed showings on the Discovery Channel, which has contracted use of the film for four years from the Lost World Balloon Society.

In 1988, Kathy commuted back and forth to Europe to earn her hot airship license. An airship is a steerable balloon. For a time, Kathy and her brother, Bill Wadsworth, used a hot air airship to advertise their company, Kat Balloons, Inc. Beside her commercial hot air airship license, Kathy is an FAA certified balloon repair person, and has been an FAA Safety Advisor and was an FAA Designated

Examiner. Through Kat balloons she is the Connecticut Sales representative for Cameron Balloons U.S. Kathy is an entrepreneur who has found a special niche in aviation.

For her 40th birthday, the two filmmakers from the South American balloon adventures created a "birthday" film from the out-takes of the miles of footage shot in Venezuela. They presented it to her as a birthday gift and called it "The Mother of All Balloonists." Everyone loved the movie and laughed until tears rolled down their cheeks. It was a great success. A sequel is possible, but the potential title unnerves the subject.

With approximately 1,000 hours of balloon flying, Kathy feels she has never had any career difficulties. "I was a draftsperson, hammered lots of nails and worked in very diversified environments inside and outside with very strong and creative personalities both men and women. I always try and learn from these experiences, situations and people. I want to ever improve my skills and capacity to produce. I want to be aware of what it takes to achieve higher levels in whatever I am doing. I've never had a 'grand plan' for my life. I just follow my instincts and so far so good."

In her spare time Kathy still does freelance drafting and site design, along with creating monoprints, drawings, and paintings in her studio on her balloon farm. She has shown some of these works in various shows throughout Connecticut.

Her happiest and most rewarding challenge these days is being the best mother possible to her son Ross, now six years old. "This is something I always wanted and hoped for - I am happy!"

Today Kathy Wadsworth feels any woman can become involved in balloon aviation. "There are no barriers to women in ballooning. She can start as a crew person or jump right into flight training and quite literally works her way up."

Her advice to other women interested in a career in aviation: "Go for it. Every door will open if you just keep going."

Soaring

Soaring is not a new sport that women have only recently discovered. As in every other field of aviation women were soaring 60 years ago, along with the men. Anne Morrow Lindbergh was the first woman to earn a glider pilot's license and that was in 1930.

Even though women have been soaring for years, the numbers of women actively engaged in the sport has remained relatively small. Today, in 1992 the Soaring Society of America lists between 500 and 700 woman as active pilots. The reasons for this in the past have been, lack of discretionary money and women acting as the primary care giver for children. In the 1990s, both of these reasons are becoming less of obstacles as more women enter the workforce and the alternative child care programs become available.

Soaring is a solitary experience. For the serious soaring pilot, four to five hours aloft is common. For many years, women soaring pilots have had an added

Fig. 5-22 Britt Floden

layer of isolation. Although the experience of soaring takes place in the cockpit, much learning takes place on the ground before and after the flight during "hangar soaring" sessions at the gliderport. Here, during these "hangar soaring" sessions gender may be a factor in limiting the number of women who might otherwise engage in the sport. Hetty Freese, who in 1972, had more than 2,000 hours in gliders said, "Quite often I have found myself the only woman in a group of men; while their wives were sitting in another group, talking about more "feminine" (?) subjects.[21]

 This to is changing. In 1985, at the women's soaring seminar held at Air Sailing Gliderport near Reno, Nevada, Nancy Evans proposed that a women's soaring group be founded. The seminar's attendees chose six pilots as a steering committee to organize the group: Sharon Smith of Texas, Nancy Evans of California, Nancy La Riviere of Washington, Janet Clark of California, Diana Doty of Ohio, and Patricia Valdata of New Jersey. So, in a camper on the sands of the Nevada desert on June 29, 1985, the Women Soaring Pilots Association was born. Its purpose, to constructively promote the participation of women in soaring.

Britt Floden - Soaring Pilot

"I am a born adventurer," says Britt Floden, "and I can't imagine what my life would be like if Charles Lindbergh hadn't flown to Paris. As a toddler I soaked up every word I heard about his flight. At the age of three, I was determined to fly the Atlantic someday."

Britt Floden was born in Sweden and like many other fledgling pilots, had some start-up problems when she entered aviation. "My father didn't understand my urge to fly, so I had to go behind his back when I was learning to fly primary gliders." That was in Stockholm, Sweden, in 1946, in the days before they had dual instruction.

"In those days, it was solo on the very first hop," said Britt. "I was full of confidence when flying alone, but scared stiff when I later flew with an instructor."

Britt's instruction went like this. They used an auto tow and the glider rolled behind the car, with Britt first learning to balance the wings. Then came the "hops." With the stick neutral the car drove a little faster and the glider lifted off and stayed a few feet off the ground before settling down as the car slowed.

"In small, safe steps we learned to take off, release, and fly straight ahead and land." The requirement for an "A" badge was five 25- second flights, and one 30-second flight.

As Britt sat ready to take the last flight for her "A" badge tragedy unfolded before her. Another instructor was accidently released from tow (probably not hooked up properly) too low, and he tried to turn back, stalled and crashed. The pilot lost a leg and the Stockholm Soaring Club lost a glider that wasn't paid for. Britt completed her "A" badge the following day.

For the "B" badge, Britt learned to make turns. The instructor stood on the ground watching and sweating. There were no instruments or a cockpit either; just a wooden seat and the wind in her face. Britt and her fellow pilots listened to the noise level of the glider to keep the right flying speed. Her test for this badge was to fly a triangular pattern and land within a fifty meter mark. She passed!

For the "C" badge they used a winch. The flight lasted three to five minutes, more if someone was lucky enough to get a thermal.

Britt got her glider pilot's license in 1947 and within a year had claimed the first female soaring record in Sweden.

It was a mid-summer's eve in 1948, when Britt set out on her quest to fly the Silver Distance (fifty kilometers) in a Grunau Baby. By this time she had logged 116 flights and eight and one half hours in the air. The map Britt had in the cockpit only covered 50 kilometers but that didn't stop her. When she ran out of map, she kept on flying, over the edge of the map and on to a record 157 kilometers away from her departure point. She won a trophy for that flight and three years later the Aero Club awarded her the official record. See Fig. 5-22.

On her five hour duration flight for the Silver badge, flying a Moswey III, Britt became violently ill and decided after four hours to land. At the last minute

Fig. 5-23 Britt Floden

a strong thermal came along and Britt changed her mind. When she landed she was only the second woman in Sweden to earn the Silver badge. Britt then married an instructor and engineer and settled down to raise two daughters. But by 1956, she was back up with the eagles.

When Britt went back to soaring she decided to earn her glider instrument rating and try for the Gold distance flight.

When she announced she was ready to try the Gold flight, the chief instructor said she could not go. He gave her various excuses but Britt felt he did not trust her.

Disappointed, Britt went off to her room, crying her heart out. When she returned to the field the chief instructor was gone, and her instructor convinced her to make the flight. "I knew this would be my last flight from Alleberg, but you can't pass up a good opportunity."

Britt took off with a map, barograph and a list of soaring clubs along the way. She carried no food and not enough money to get back home.

The flight went perfect, until the landing. Britt landed her glider at the end of her flight in a sugar beet field. The farmer threatened to sue her for the damage to his crops. While the farmer's threat had her concerned, she was elated because she had set a feminine Swedish distance record of 305 kilometers. That record still stands in 1992.

Britt wrote an article about the incident for the Aero Club. They paid her 100 crowns for it, and that settled the score with the farmer.

Britt completed her Gold badge in 1958, flying a Weihe and used her recently acquired instrument training to fly in a thunderstorm. In doing so, Britt set an altitude record of 3,365 meters and became the first woman in a Scandinavian country to earn a Gold badge.

In 1960, Britt finally got to fly across the Atlantic - to the United States - her new home. When she and her husband emigrated she brought with her three distance records, an altitude record and two daughters.

Britt again put soaring on the back burner again for about 10 years. "I missed it very much," she recalls. "And I was delighted to get back into the air again, this time at Van Sant Airport, in Pennsylvania." Once in the air, Britt began to stalk the records again. She quickly acquired three national feminine records. In addition to her Swedish records, Britt Floden's records include three U.S. Feminine speed records.

Britt also went back to Europe in 1973, and in 1975, to compete in the Women's Internationals in Poland. There she earned her first Diamond (three are needed for a full Diamond badge). "I think I would refer to these contests as learning experiences," she said enigmatically.

Britt's appetite for adventure appeared unrelenting. She went on to fly hot air balloons, and earn her commercial balloon rating, and build a BD-4 airplane with her husband Bjorn. Today Britt is retired from soaring and has become an accomplished tapestry weaver.

Virginia Schweizer - Soaring Pilot

Virginia Schweizer started flying gliders right in her own backyard, taking instruction in a vacant field behind her father's potato farm in New York at the age of 18. She took the student's place inside the single place trainer and listened to the instructor who sat outside on the back of the tow cart to explain the maneuvers before each flight. As a modern day teenager, Virginia marveled at the shear adventure of doing something "different" like gliding but was taken by the solitude and gracefulness of sailplaning. She continued her training and become the first woman in the United States to earn a commercial glider rating. See Fig. 5-24.

Virginia participated in competition gliding and set women's distance and cross country records including the coveted Silver Caward in 1946. The first ever 2 hr flight covered 38 miles. Recounting how advanced gliders have become since that era, Virginia laughs and remarks that with today's high performance ships, a flight of that distance could be completed in a fraction of the time, just 10 minutes!"

After World War II, Virginia provided glider training in army surplus two place trainers and inspired men and women alike to take up gliding. Her active involvement during that period became a part of a 10 year effort to win the FAA's approval to adopt an aircraft towing method which would change gliding forever. "After the war we used PT-19's and Stearmans as tow planes but it took years to convince the FAA on the merits of that approach."

Logging more than 1,000 hours riding thermals, Virginia admits that glider flying is an indescribable feeling and the enthusiasm is contagious as she recounts the joy of soaring.

"I can't explain just how wonderful it is, you have to experience it yourself to understand it but I strongly believe in starting at an early age. If you start when you're young, as I did, and make it a major portion of your life, it is something that stays with you forever, it gets in your blood and you'll never lose the touch.

Virginia's husband is a part of the Federation Aeronautique and together they travel the globe to attend international conferences to promote aviation interests and gliding. Their home is just a sailplane ride away from Harris Hill, the Soaring Capitol of the Country, where the site of the National Soaring Museum is also located. Having spent a lifetime in the realm of gliding, the Schweizers have designed a commemorative exhibit which is on display detailing the history and progression of the sport they so dearly love.

In 1989, the Women's Soaring Pilots of America and the I-26 Association established the Virginia M. Schweizer Competition Trophy for the woman pilot with the highest score during the annual I-26 championships. This trophy is named in honor of America's pioneer pilot who was a national champion when the country held feminine competitions.

Fig. 5-24 Virginia Schweizer

Chapter Six

Entrepreneurs

Have confidence in yourself and tell yourself 'you can' twice for every time you are told 'you can't.' Confidence that you can succeed is everything. Take every negative remark as a challenge to achieve more and progress to newer heights. You are able to do anything you believe you can do. You might even surprise yourself. Alinda Wikert

Alinda Wikert - CEO Express One

Alinda Wikert didn't start out with a dream of being a pilot. In fact, there were no pilots in her family background and she didn't take her first lesson in a Piper Tomahawk until June of 1979. Her husband owned a Learjet and she would often accompany him on business trips. During the trips she would watch the co-pilot. One day it occurred to her that it didn't look hard to fly a plane, and, she thought to herself, "I think I can do that!" What struck her also was the fact that most of the co-pilots she watched were young, around the age of 21-23. It also seemed to her that she was at least as intelligent as the person in the right seat; and her judgement was better. The last thought was not subjective but based on fact. Insurance companies charged 21-25-year-old men more insurance than others. So why shouldn't she learn to fly? See Fig. 6-1

Once she earned her private rating, Alinda wasted no time advancing her skills. Within five months she had her commercial and multi-engine ratings. By June of 1980, she had purchased Jet East, at Dallas-Love Field, in Dallas, an FBO and Learjet sales, charter, and service center.

Alinda realized that if she were running a company that operated Learjets she should obtain that rating also. That's when she ran into the first problem. She was attending a flight school in Tucson, Arizona, with a fellow student, also a female, when the director of the school told her that she would not be able to

Fig. 6-1 Alinda Wikert

achieve the Learjet Type Rating she needed. When she asked why, the director told her that she didn't have enough flight time. Wikert only had 380 hours flight time and he said that was not enough to transition to jets. Wikert's fellow student was told she was too young. (She was 21 years old.) Both women knew that the information was wrong and that the military often transitions their new pilots to jets with 70-80 hours, and seldom do their pilots have as many as 380 hours. With such opposition, Wikert and her fellow student chose to take their check rides for their Learjet Type Rating in Dallas from the FAA. They both successfully received their type ratings on the first ride in October 1980, one year after Wikert received her private pilot rating.

From that point on it was full speed ahead for Alinda Wikert. She immediately began flying charters in the Learjet and in 1984, added two more ratings to her growing list. She earned her rotorcraft rating and became a Whirly-Girl, and after attaining the necessary 1,500 hours earned an ATP in the Citation, followed by the ATP Learjet and rotorcraft.

Along the way Alinda began to formulate an idea to help other women who wanted to fly, since it is more difficult for a woman to get a job flying if she has low time. Alinda's idea was to start a regional overnight package service, that employed only women, thus giving them a chance to increase their hours and then move on to the high paying airline jobs. When Wikert went to the local banks, they were not as enthusiastic. Who ever heard of an all-woman airline? Certainly not any of the male bankers. They didn't think the all-woman concept would "fly" anyway. So, Alinda put the idea on the back burner and went on to other things. She upgraded to first officer on a Boeing 727 with Jet East and became the first woman to fly a helicopter as a stunt double for a movie. From this "first" she was included in a permanent exhibit of Women in Vertical Flight in the National Air and Space Museum.

In 1987, Alinda sold Jet East and spun off a Part 121 air carrier operation with Boeing 727s, 737s and McDonald Douglas DC-9s. Today that operation has 18 aircraft, 100 pilots and 35 flight attendants, and is called *Express One*. And, Alinda Wikert is the only female owner and CEO of an airline in the world. She also works as a wife and mother of four children (a set of twins).

Alinda has had many helpful people in her life that guided her through her ratings and difficulties. Many of them were men. In fact she makes the point of letting people know that 90 percent of her dealings with men in the aviation community were supportive and gives a lot of credit to all the pilots who helped her along the way. She also cautions women who are thinking of becoming aviators, not to believe all that they are told. "Have confidence in yourself and tell yourself 'you can' twice for every time you are told 'you can't.' Confidence that you can succeed is everything. Take every negative remark as a challenge to achieve more and progress to newer heights. You are able to do anything you believe you can do. You might even surprise yourself."

Wanda Whitsitt - Lifeline Pilots

In 1979, Wanda Whitsitt took her first small plane ride. Her husband had always wanted to learn to fly and Wanda thought it would be a nice activity the two of them could do together. At the time, Wanda had a full daily agenda, too. As a mother of four, she maintained a house, and volunteered for the Red Cross, PTA and Girl Scouts.

It only took a few lessons for the flying bug to bite. "Before long," Wanda says, "I got hooked." She earned her private rating in 1979, followed by her commercial instrument rating, and ground school instructor certificate. Today Wanda has over 1,500 hours.

Something happened to Wanda while she was flying. She began to think about the expense of flying and how she could justify it. While she loved flying, she felt guilty over spending the money on "just herself." She began to think about the young blind man losing an opportunity to become sighted because there wasn't enough time, or the toddler losing her life because there was not enough time to get her an organ transplant. She thought about all the people who lose an opportunity to improve the quality of their lives just because there wasn't enough time. Wanda decided she wanted to put her new-found skill to use helping others. Wanda checked around and discovered pilots in neighboring states volunteered to fly blood and plasma to remote locations. "We don't have any really remote locations here in Illinois," she said, "so I contacted several local and state social service agencies. I discovered there was a need for pilots to transport patients to hospitals in other cities and states. See Fig. 6-2.

Wanda Whitsitt's idea quickly began to take a real form. As she talked about it among her pilot friends she found she had more volunteers and support than she ever dreamed of. Even today there are more volunteers than missions.

Wanda had no trouble coming up with a name for the organization. "Lifeline Pilots," she said. "Because we extend a lifeline to people in need."

Over the years Lifeline Pilots has grown and Wanda has nurtured its growth every step of the way. She has recruited pilots, coordinated missions, raised funds, written a newsletter, handled public relations and flown the missions herself.

At first Lifeline flew only missions within the borders of the state, handling half a dozen missions a year. In 1989, Lifeline flew 57 missions and extended its "lifeline" to as far as the aircraft would fly, usually about 500 miles, or about a tank of gas.

Lifeline pilots are prepared to fly organs, blood and blood products and other tissue, as well as medical teams, patients who are mobile, disaster relief supplies and personnel.

Wanda has recently been instrumental in helping to form Air Care Alliance - US, an organization of 25-30 groups like Lifeline. Their intention is to fly relays and extend their helping hand across the country. Wanda is working on getting an 800 number so she can coordinate the cross-country flights.

Pilots of Wanda Whitsitt's Lifeline have flown more than 550 missions and because of the growing publicity, they now average 15 missions a month. The pilots also have the same sharing attitude and generosity as Wanda. They do not get paid for their time and they pay for their own fuel and maintenance costs. Although the Lifeline Charter allows for a pilot to be reimbursed 25 percent of his or her fuel costs, Wanda can't remember anyone ever asking for the money. Wanda estimates that if Lifeline charged for these missions it would cost from $300 to $600 a mission. That fee includes the pilot's time, value of the aircraft, fuel and maintenance expenses. Funding for Lifeline's missions comes from

Fig. 6-2 Wanda Whitsitt

private donations and word-of-mouth advertising. Wanda's volunteer group now includes more than 300 pilots in 10 states.

As much as Wanda likes to talk about Lifeline, she does not brag. If pressed, she will mention its perfect safety record, and quietly acknowledge that in 1986 she was inducted into the Illinois Aviation Hall of Fame for her work in organizing Lifeline. In 1992, she received a Certificate of Honor from the National Aeronautical Association at a ceremony at the Smithsonian Institution in Washington.

But Wanda will not hesitate to articulate the goals of Lifeline. "Responding to everybody in need is our biggest goal."

When Wanda talks about the future for women in aviation she has a good frame of reference. "Fourteen years ago, when I began taking lessons, I remember a woman pilot at an airport shocked some people. Not any more. We are judged on our ability. That's the way it should be.

"Today we are often caught up in negative messages. They are all over the media. Flying provides a positive influence and I have met and been encouraged by many generous and compassionate people. I have had the opportunity to know and work with some of the finest people I have ever met."

Anne Gray - AirLifeLine

Anne Gray worked on Wall Street where she earned a good salary but worked up to 14 hours a day. Today that is all behind her even though she loved her work and it was rewarding. Her husband worked a similar schedule on Wall Street. It was a tough grind and they decided that one person on that kind of schedule was enough in the family. See Fig. 6-3.

After she quit, Anne found staying home did not meet her expectations. One day she drove to the Somerset, New Jersey airport near her home. On her first flight she was hooked, but she got motion sickness and that put a damper on things. "There was no way I was going to let it keep me down," she said. "I forced myself to go to the airport for each lesson, until I had complete control of the airplane, which eliminated the problem." After she earned her private rating, Anne moved along to earn her instrument, commercial and multi-engine ratings. Anne also takes recurrent training in Wichita, Kansas, with FlightSafety International, where training and standards are designed primarily for professional pilots. Today she is just a few hours from her ATP. She has over 1,450 hours in her log book (over 1,250 as Pilot-in-Command).

Anne was in love with flying but feeling a little guilty about not sharing it constructively with others. She began to wonder how she could use this new found skill to benefit her community. Then she heard about AirLifeLine at an AOPA seminar. Anne learned about their work and that they needed qualified pilots.

Anne joined AirLifeLine, a nation-wide organization based in Sacramento, California, and another group, the Volunteer Pilots Association, of Pittsburgh. Anne, like the other volunteer pilots in both the organizations, donates her time, the use of her aircraft, and the cost of the fuel and maintenance of the aircraft. In August 1992, AirLifeLine had 728 volunteer pilots in 49 states. This year they will fly approximately 1,300 missions for a total of 1,150,000 miles. Anne has answered the call over 55 times since she joined the organization. She might be called to pick up a human organ and fly it to a patient awaiting a life-saving transplant. Or she might get a call to fly a patient with a life-threatening disease to a distant hospital for treatment.

Time is a critical factor. That's why general aviation pilots are often called upon to make such flights. Volunteer pilots like Anne are available around-the-clock, and they are able to fly into thousands of airports not served by scheduled flights.

Anne is on call 24-hours a day. When Anne is called she flies her twin-engine Beech Baron. Anne first started flying patients in a single-engine Piper Dakota but soon traded up for the Baron. Now she will get flights that other pilots cannot take. The Baron is equipped with de-icer boots and alcohol for the props and windshield. It has both weather radar and a stormscope. That makes it possible for Anne to take flights that might ground other aircraft.

Anne made the first of her AirLifeLine flights on January 15, 1990. She flew a three-year-old girl afflicted with Rhetts Syndrome from Teterboro Airport to Baltimore-Washington International Airport. The little girl was going to Johns Hopkins for treatment.

Fig. 6-3 Anne Gray

She was accompanied by her mother and social worker. When they saw the "little" single-engine airplane, they were hesitant. And when they saw the pilot was a woman, they were even more hesitant. But they said, "What choice do we have?"

After the flight, Anne got a note that read, "It is so wonderful that there are people like you who are willing to help those in need."

When the little girl needed transportation again, to Philadelphia, they requested Anne.

All of the patients Anne flies for AirLifeLine are in financial need. Their finances have been depleted by medical bills. All are appreciative. They often send her cards, notes and photographs afterward, thanking Anne for the service.

Anne flew a nine-year-old boy from Syracuse, N.Y. to Rockefeller University Hospital, in Manhattan, several times. He has a rare disease that prevents his skin from growing with the rest of his body. He requires periodic special skin graphs and while the procedure is not long, ordinary transportation would keep him out of school for days.

Anne carries a beeper so that she can be reached anytime, anywhere. Once she was attending a dinner near Republic Airport, on Long Island. Her beeper went off. It was an organ flight from Elmira, N.Y., to Pittsburgh. The call came from the Volunteer Pilots Association. Most of their calls are for organ transport.

Anne flew back to Somerset Airport to get her Piper Dakota. She picked up a co-pilot and flew the mission. Two pilots are required on all night flights.

Anne files lifeguard flight plans when she's carrying a patient or an organ, and air traffic controllers give her preferred treatment, providing direct routing whenever possible.

While Anne quit a high pressure 14-hour-a-day-job, she still works long hours sometimes. Her volunteer flights frequently require her to put in 12 hours or more a day. She may have to fly an hour or two to pick up a patient, then fly them to the hospital, wait all day while they undergo treatment, fly the patient back home, and then fly herself back to her home base.

On one mission, Anne was on a final approach into Teterboro to pick up a patient when the cockpit filled with a beeping sound.

"For a moment I was concerned," she said. "Was my gear down? Was it a stall warning? What was wrong? It was scary."

It was Anne's beeper calling. Her coordinator was setting her up for another mission before she had finished the one she was on.

In December 1991, Anne had a passenger-patient born five weeks earlier in Erie, Pa. The infant had a heart defect that until 1979 had always been fatal. The Children's Hospital in Philadelphia had developed a three-operation procedure that could save most of the youngsters born with this defect.

A week after his birth, the little boy underwent open heart surgery. His father took time off from his job to stay with his baby son.

Leo's mother, a nurse, stayed home with their three-year-old daughter, while recovering. But she managed to fly back and forth to Philadelphia with Anne to see her baby. The note Anne received said, "Thank you for making our Christmas wish come true."

She and her daughter wanted desperately to have the baby and his father home for Christmas. Anne helped make that possible.

"Flying is a passion for me," she says, "and it's very rewarding to do something I love and help people at the same time."

Her husband, who is not a pilot, is very proud and supportive too. When an article about Anne and AirLifeLine appeared in a New Jersey newspaper, he took the article to work and showed it around. He raised $9,000 in donations for the organization.

Anne is highly motivated and committed to her volunteer work. "One cancer patient told me he would have died were it not for the treatment he received at the hospital I was flying him to. That makes it all worth while."

Cecile Hatfield - Aviation Attorney

Any woman who starts flying lessons with a newborn son is not afraid of challenges. Cecile Hatfield is such a woman. Cecile started flying in 1962, just two months after the birth of her son. Cecile would pack her young son Andy and her two-year-old daughter, Laurie, off to the airport whenever she took her lessons. What sparked her interest in flying was a brother and sister (she has two brothers and two sisters) who had taken flying lessons. "Flying was something I had always wanted to do," she said. See Fig. 6-4.

But flying for Cecile was easier said than done. When Cecile drove to Opa Locka airport for her first lesson she got cold feet. "I turned around three times to go home," she said. But Cecile was determined to fly!

Cecile's quest for wings had other roadblocks along the way. The next challenge was facing her own fear. "I finally got the nerve to go to the airport and arrange for my first lesson. I actually got into the plane and the instructor took off." Then her real problems began. "I was terrified. I prayed, and my mother prayed, too." Once Cecile had "grounded" her fear she had a new challenge to face. "I got vertigo and sick, very sick. I usually didn't get sick until we were on the ground but the vertigo was awful."

She had started out in a seaplane and her instructor suggested that she switch to a land plane and fly over the ocean or over land early in the morning or late in the afternoon. That helped cure Cecile's vertigo and she built up a tolerance to the bumpy air. From then on it was full bore ahead to her private license.

Cecile is not sure when she was hooked, but she said, "I realized flying was easier than driving. There were two or three times that I was almost killed driving to the airport."

Cecile enjoyed flying more than anything else, and within a year she had her private license. She followed that with her ground school instructor's certificate.

She then joined the Ninety-Nines and got involved with the Florida Women's Pilot Association. At the time, the Florida Women's Pilot Association was sponsoring an annual intercontinental air race called the Angel Derby. As many as 100 planes flown by women would span Canada, the United States and Mexico. Cecile thought becoming involved in that would be fun. It was. The race went a long way toward promoting women as competent and safe pilots, and the women also served as unofficial goodwill ambassadors. Before each flight the women would stop at the White House and pick up "Letters of Friendship" from the president or vice president. During these years Cecile got to meet vice presidents Spiro Agnew and Hubert Humphrey. Humphrey even asked Cecile to be his personal pilot.

Cecile Hatfield flew her first air race in 1964 and flew in three other Angel Derby races.

Cecile likes getting involved and before long she was organizing and conducting the race. She did this from 1966 to 1972. She was race chair for a year, in 1967, and president of the association from 1970 to 1972. The Angel Derby races ended a few years ago, but not before women like Cecile Hatfield chalked up more than 50 years of showing America that ordinary women could be pilots and have fun at it too.

A strange thing happened to Cecile along the way. She was having so much fun she began to feel guilty. "I should be doing something more constructive," she said. Although she had a degree in education, Cecile had been interested in the law ever since she had been an undergraduate at the University of Florida ten years earlier. She had thought that someday it would be nice to be a lawyer. Well, that day was fast approaching, although Cecile perhaps didn't see it coming. When she thought about a career in law, she saw the long range possibility of becoming a lawyer and perhaps doing something around aviation. If she could combine the two then she could have fun and help people.

With her decision to go to law school, Cecile was setting up a future that she would never have predicted. In 1975, Cecile received a Juris Doctorate from the University of Miami School of Law. It is fitting to note that she made the Dean's Honor List and also graduated in the top 15 percent of her class.

Between her undergraduate degree and her law degree, Cecile worked at several jobs to support her family and her flying. She did TV commercials, starred in a movie with Frank Sinatra, filmed in Miami Beach, and worked as a fashion model for an Italian designer. "Whenever I got a modeling job, I took some of the money and went out to the airport for a lesson or to keep current," she said.

With her law degree in hand, Cecile was about to break new ground for women in the fascinating field of aviation law. Cecile Hatfield became the first woman aviation lawyer who served as a trial attorney for the United States Justice Department, the first woman to serve as an associate general counsel for the Piper Aircraft Corporation, and the first woman to chair the Aviation and Space Law Committee of the American Bar Association Section of Torts and Insurance Practice.

Cecile served nine years as a trial lawyer with the Department of Justice in Washington. She was the lead trial counsel responsible for national trial practices involving complex air carrier, military, and general aviation, as well as admiralty and appellate litigation in federal courts. She also supervised local counsel and client agency attorneys in defense of the Federal Aviation Administration, National Weather Service, Department of Commerce, the National Oceanographic and Atmospheric Association, the Department of Interior, Department of Agriculture, Coast Guard, U.S. Army, Navy and Air Force.

After leaving federal service, Cecile served as head of the aviation law department of Peters, Pickles and Neimoeller and later as a trial attorney for the firm of McDonald and McDonald where she was involved in cases of aviation law and insurance. She also has defended pilots, manufacturers and the airlines. "When I defend a client it is my responsibility to keep the costs of litigation down. If the verdict happens to be against my client I try to keep it so my client is not forced out of business."

On January 1, 1991, Cecile opened her own law firm, the Law Offices of Cecile Hatfield, in Miami. A year later, with a growing case load, she hired her first attorney, a former Eastern Airlines and helicopter pilot.

Cecile also lectures and writes articles on aviation litigation for the American Bar Association, the Lawyer Pilot Bar Association, Embry-Riddle Aeronautical University and the Lloyds of London Press Aviation Symposium.

Fig. 6-4 Cecile Hatfield

In April 1992 she conducted an Aviation Law Symposium in Brussels. Thirty-four countries attended.

Cecile is certainly a woman who loves aviation and is contributing everyday to enhancing the lives of the people around her. Cecile is a role model to show young women that anything is possible with determination.

"Go for it," she says. "There is nothing to hold you back if you have faith in yourself, determination and a little discipline."

Betty McNabb - Past International President - Ninety-Nines

Although she didn't learn to fly until 1951, when she was 42, Betty McNabb has deep roots in aviation. Those roots go all the way back to December 17, 1903 at Kitty Hawk, N.C. Betty's father, Frank Wood was present and helped with the mechanical adjustments of the Wright brothers' *Flyer* and was one of the few people on the beach watching the first human to successfully fly a heavier-than-air machine.

Betty decided to fly while working as a health records consultant in Georgia. She was working for the State of Georgia at the time and the roads were terrible in those days. Her job required her to be out and on the road at 5 a.m. and often she didn't get home until 7 p.m. "I had to drive, which I never liked, and the roads were red clay, and I got flat tires at night when I was all alone. I didn't like that at all. I began to think I must either learn to fly or give up this job, which I was enjoying very much. So, I learned to fly.

"My husband would have had a fit if he had known I was learning to fly, so I didn't tell him until I soloed. It was my money, I mean I was working. When he found out I had spent so much, he said, 'go ahead, finish.'

"I was scared for the first two years I flew," said Betty. "When my instructor said 'You're ready to fly,' I said oh, no, not yet. I made him let me have two more hours flying practice. Then when I got in the plane to solo, I said to myself, 'Well, let's just pretend that John is here.' So, while I was in the air, I talked as if he were sitting next to me. When I thought I had a problem I would say, 'John, what do I do now?' Then I 'thought' what he would have told me." Betty passed with flying colors.

"Also, during the first two years I didn't realize that if you treat the airplane the way it should be treated, it will treat you all right. Ever since that dawned on me I've been doing that ever since and loving it."

It was her original instructor who inadvertently motivated her to go and get her ATP rating. She had bought an Ercoupe on her instructor's recommendation and one day had just returned in it from a consulting trip.

She jumped out of the plane and walked over to her instructor who at the moment had his back to her talking to another student. The student said to him, "John, could you find me an Ercoupe like Betty's?"

John replied, "No, you don't want that kind of an airplane. It's engineered for three kinds of people, the very elderly, the handicapped folks or morons."

Betty thought to herself, "I was not that old, nor handicapped so,??? At that moment I decided to go all the way."

Betty and her husband moved to Florida, and each time she completed a rating, she let John know. He would reply with his congratulations. "When I completed my ATP he sent me a big card with huge red letters saying, 'WOW'!!! Bless John, he really gave me incentive."

Betty went on to obtain multi engine, sea, glider, CFI-I and ATP ratings. She owned seven different airplanes and did ferry, charter and general testing. See Fig. 6-5.

In June 1958, the commander of the Air Force base in her home town called her. "Hey Betty," he said, "would you like to fly through the sound barrier?"

She thought he was joking and replied, "Yes, and when are we going up to the moon?"

It took the Air Force officer several minutes to convince Betty that the offer was genuine. The general knew she had been through the hyperbaric chamber and passed with flying colors.

"At the base they took me to the equipment room where they have the flight suits. Being a woman and well-endowed, I was not shaped properly to fit into one of their suits. Two young pilots weak from trying not to break into laughter stood me up facing against the wall. Finally the flight suit and I were one. We

Fig. 6-5 Betty McNabb

laughed long and hard over that and it only helped make the day even happier. I also vowed to lose those few pounds after that episode, which I did.

"It was a wonderful experience. My pilot let me fly the F-100F Super Saber jet through the sound barrier. I was very excited to be the eighth woman in the Western Hemisphere to have done this."

As soon as Betty soloed she began making her business trips by plane. She even flew from Atlanta, Georgia to Chicago in her Ercoupe. For this flight she admits, a little reluctantly, she went "IFR," (I follow railroads).

Over the years, Betty put her airplane down in every state except Alaska and Hawaii. "Medical records work was just beginning to grow in those days and I had to fly all over the United States.

"One time I stopped in Arkansas at a tiny little dirt airstrip. The wind was blowing strong against me, so I decided to land and refuel. I walked into a little, old building and surprised an old man in there. I asked him for some gas and he said, 'mam, we don't sell car gas at all.' I said, 'I don't want car gas, I want fuel for my airplane.' He said, 'You do?' He looked at me and then said, 'well, all right, but your pilot will have to sign for it.' Then I said, 'all right, here she is - I'm the pilot." He looked at me in disbelief and I said to him, 'go out there and see if you see anybody else out there. He did, and then came back much embarrassed. 'I just never seen a lady pilot,' he said weakly. 'I'm going to remember you!' He apologized again and again and said, 'Here's your gas - you don't have to pay for it.'"

Over the years Betty has been the recipient of many awards and honors for her distinguished aviation career. As Lt. Col. Betty McNabb, she was the first female officer in the Air Force to graduate from the War College in 1964.

Betty was also served as a senior member of the Civil Air Patrol, and was inducted into the national aerospace Education Association Hall of Honor in 1977.

Betty served for more than 22 years (from June 1969 to June 1991) as an active Coast Guard Auxiliary and senior pilot. Of the more than 9,000 hours she accumulated as a pilot, more than 1,100 of these hours were flown on Coast Guard missions. She has the distinction of being the oldest active pilot in the Coast Guard Auxiliary and was still flying at the age of 82.

On June 21, 1991, the Coast Guard presented Betty Mcnabb with the Award of Merit in recognition of sustained superior professional leadership in support of the Coast Guard Auxiliary. The award read in part, "...she has provided years of sage advise to generations of auxiliary pilots. Her exceptional personal dedication, judgement and devotion to duty are most heartily commended and are keeping with the highest traditions of the United States Coast Guard."

A new award was devised in 1990 by the District VIII Commodore (Coast Guard Auxiliary) named the "Betty McNabb Aviation Award." Betty has been given the honor of structuring and presenting this award.

Betty successfully combined an aviation career and a medical records career into organizing and directing aerospace education seminars and medical records seminars. She has spoken at many colleges and universities and each year for the last 34 consecutive years (as of 6-11-92) has given a lecture at Middle Tennessee State University.

Over the more than 40 years Betty has been flying, she has had the opportunity to become involved in many aviation organizations. She is President Emeritus of the Ninety-Nines, after serving two terms as their president, she is the past vice president of the American Yankee Association, (Grumman Aircraft Association owners organization) and she is a member of the United Flying Octogenarians. She became eligible to join this organization on August 5, 1989. The requirement of this organization is that the person be over 80 years of age and still flying. As you can imagine, Betty is quite proud of this particular accomplishment.

Amy Carmien - President/Publisher *Women in Aviation*

Amy's initial memory of aviation was a short story in her first grade reader. "It was a story, about Amelia Earhart, and it sparked an interest somewhere in

Fig. 6-6 Amy Carmien

my seven-year old mind." Amy began looking for other books about Amelia. No one in Amy's family flew, but the seeds had been planted nevertheless. "Somewhere around the seventh grade it occurred to me that I might be able to fly airplanes too."

Amy and her family lived about fifteen miles from an airport and she searched for someone to take her for a ride. As luck would have it, Amy's eighth grade teacher had just received her private pilot's license. She was eager to share her new-found pleasure.

Amy's first flight bonded her fascination with flight permanently to her being. She took her first flight on her sixteenth birthday and Amy's supportive parents drove her to the airport for her flight lessons. Amy soloed before she had her driver's license, and had earned her private pilot's license within a year. Her first passenger was her father, followed by the woman who had introduced her to the air and shared her flight training manuals with her.

The fascination about flight did not ebb in Amy. She flew as often as she could and she soon was taking an engine class in college. It was then she realized how much there was to aviation and how much more she wanted to learn. She transferred to Embry-Riddle University's aviation maintenance technology program with the idea of earning her airframe and powerplant license. When the school offered her a maintenance fellowship, Amy decided to stay for her bachelor's degree. During that time she interviewed for the NTSB as a co-op program student and passed. She then spent two semesters in their program investigating aircraft accidents from their Washington, DC office. See Fig. 6-6.

After graduation Amy worked with the NTSB in their Chicago office for about a year, then decided to move back home to Traverse City, Michigan. She secured a job as an A&P mechanic and during this transition, the idea for a publication for women in aviation began to form. "I had always had a strong writing/journalism interest and I decided to combine it with aviation."

In September 1988, Amy began the business planning for a publication that would become known as *Women in Aviation*.

"Throughout my aeronautical career I have seen a sore lack of recognition for women involved in aviation. Other than the occasional 'aviatrix' article in the local newspapers, it seemed to me that women in aeronautical fields were for the most part invisible, isolated individuals. I began to wonder why a publication covering women in *all* aviation fields did not exist."

Amy wrote letters, made phone calls and had many conversations with women in aviation fields. And women responded with enthusiasm. "Soon I began to realize that it was time to create a new dynamic newsletter written by women, about women and for women in aviation.

The primary purpose of *Women in Aviation* is to acknowledge contributions from women in all sectors of aviation, regardless of their flight status. In the more than four years that Amy's publication has been in existence, it has been growing and covering a wide variety of topics, including meteorologists, balloonists, mechanics, airline pilots and aeronautical engineers.

Women responded to Amy's initial queries for publication and demonstrated an enthusiasm that has helped make the publication a success. "Women who did not know me, generously open their homes and telephone lines, loaned photographs and rearranged their schedules to make the first issue possible." Today Amy has the continued and enthusiastic support from such prestigious organizations as the International Society of Women Airline Pilots, and The International Women's Air and Space Museum.

"I think it demonstrates one thing about women in aviation. We are willing to help and learn from each other. Although we may have been a little ahead of our time in 1988, the interest in women in aviation has certainly caught up with us and surged ahead. The multitude of books now being published on the subject confirm that. The Women in Aviation Conferences, now in their fourth year, the International Women's Air and Space Museum and our magazine are all experiencing growth. The network between all these individuals and events is incredible.

"As a publication, we will continue to keep the focus on individual women in the various fields of aviation. By documenting the achievements of today and providing a glimpse of the past, we hope to preserve the aviation sector of women's history. With our history to build on, the next generation of women in aviation will be a mach or two ahead at the outset. And if one single woman in aviation feels a bit less isolated because of this publication or these events, then we are nudging the process in the right direction."

Racquel McNeil - A Ladybird

Racquel McNeil is probably typical of many women in aviation. She didn't own an FBO and has not set a world record (not yet anyway). She just always wanted to fly. One might even call her a "late bloomer" when it came to aviation. Raising six children consumed her time so she put flying on a back burner. Then one day, on her 40th birthday, she got a surprise. Her husband presented her with a birthday gift, a certificate for a Cessna "Discovery Flight." This was the motivation she needed to pursue her dream. "It was one of the best birthday gifts I ever received and my husband still jokes about still paying for it."

Racquel found the time between raising six children to take her flight lessons. "It took me a little bit longer than others, because I couldn't get out to the airport. I had to work around the schedule of my family." See Fig. 6-7.

Racquel's most vivid memory of flying is the day she soloed. "One day, my instructor, Janis Blackburn said, 'Okay, take it around the pattern yourself.' I was scared to death," she said. "I thought of holding on to Janis' legs, and not letting her out of the airplane. However, I knew I had to do this thing myself. I had to let go, and spread my wings. It worked out just fine. I look back now and realize that where there is a will, there is a way."

After Racquel got her private license she went out and bought a Cessna 150. It wasn't long before she yearned to improve her pilot skills, so she decided to

Fig. 6-7 Racquel McNeil and her greatest supporter, husband Joe.

take instrument instruction. It turned out that the plane at the airport did not have the instruments for instruction. Racquel was not deterred, she went out, bought the instruments and had them installed in her Cessna. After Racquel earned her instrument rating, she flew the Cessna for about five years. "It's funny," she said, "how perceptions are slow to change. Whenever we landed at an airport the ramp person always came up to my husband and ask if he wanted the tank topped off. Even today his reply is , 'ask her, she's the pilot.'"

Racquel's children grew up, and soon she had three grandchildren. This required Racquel to make a small change in her flying life. She sold the Cessna and went out and bought a larger plane. "I went out and bought a Piper *Warrior*," she said. "After all, the Cessna was too small to fit all the grandchildren."

When Racquel started flying the Piper, she began to feel different too. "I was now able to take long trips to visit my married children, first to New Hampshire, and then to North Carolina. I also felt more professional.

"When I look into the future, I see no reason any woman can't learn to fly. There is no special strength involved. And there is a special freedom that flying gives you. It's fun and empowering - and there's nothing safer, if you are careful and tend to all details.

"The future for women is wide open," says Racquel, "from a Cessna to the Space Shuttle - it's ability that will move you ahead."

Moya Lear - Lear Aircraft

It is said that "behind every great man is a great women,"and that phrase certainly rings true for Bill and Moya Lear. Moya Olsen Lear was born in 1915 and grew up in the vaudeville days as the daughter of Ole Olsen, of the renowned Olsen and Johnson comedy team. She married the flamboyant inventor, Bill Lear in 1942 and began a 36 year career as dedicated wife, mother and confidant to her genius husband. Though his fortunes sometimes swung from vast wealth to the edge of bankruptcy, depending on the success of his latest invention, Moya was always by his side sharing in his discouragements as well as his spectacular triumphs.

As one of our countries great inventors, Bill Lear's successful inventions include: the first practical car radio, the eight-track stereo, the aircraft ADF, (automatic direction finder,) and the Learjet. Tragically, when he died of leukemia in 1978, the design of his last great project, the revolutionary Lear Fan aircraft, was still on the drawing board.

The Lear Fan would be the safe, economical, lightweight and fuel efficient aircraft the ailing general aviation industry had been looking for. Bill's design was a pusher, propeller driven by twin turboshaft engines, with jet like speed, yet far more economical. By using an innovative graphite epoxy composite material, its fuselage would be stronger than aluminum of the same thickness, but at half the weight. When Bill proposed that the entire plane be constructed from this material, industry specialists were skeptical.

Fig. 6-8 Moya Lear

Convinced the Lear Fan idea would "fly", Bill immersed himself in the planning and engineering of the plane until his final days. His dying wish was for Moya to "finish it" and make sure his last great design would get off the ground. Although the prospect of going on without the man she had loved for 40 years was more than she could imagine, she remembered her promise." After he passed away, I went to the plant and sat down at his desk, in his chair, and I felt him there and heard his instructions: 'Mom, be sure and finish it!'" See Fig. 6-8.

Though neither an engineer or a businesswoman, Moya courageously stepped into the role of Chairman of the Board of the LearAvia Corp. She had received her formal education from Ohio State University, Pace Institute of New York and, University of Geneva, but spent most of her life as a wife and mother to Bill and her four children. Moya had no practical work experience to prepare her for the monumental task ahead as CEO of an aircraft corporation. "I had always been caught up in my husband's work and learned from him through osmosis. I simply carried on to the best of my ability, surrounding myself at the plant with dedicated and knowledgeable people."

Never lacking in her endeavor, Moya had a gift for galvanizing the talents and energies of her loyal employees and in turn they rallied to her support. As a motivating force, she piloted her team to the prototype completion and ultimately to the inaugural flight of her husband's Lear Fan aircraft on December 32, 1980. (In an agreement with Lear Fan to provide funding for the project, the aircraft's first flight had to occur before the end of the year. To keep the agreement, the British government declared the day of the flight, December 32, 1980. Envelopes carried aboard the first flight were canceled with this date, and the United States Post Office honored the cancellation.)

She had kept her promise and "finished it" for Bill, within three years of his death. It had been a labor of love, and her dedication to his dream had elevated her to entrepreneurial success. Recounting modestly, "They all tell me that it wouldn't have been possible without me but I have to think that it was such a great organization they would have done it anyway."

The Lear Fan was near certification when the project culminated in 1985, with three aircraft involved in flight testing in Reno, Nevada. The industry was no longer skeptical about the use of advanced composites. The validity of Lear's revolutionary ideas continues today, as aircraft manufacturers and designers incorporate this material in many ways on their products.

Widely heralded as an aviation sensation, Moya has attained a long list of honorary degrees and humanitarian titles. She was named an Honorary Doctor of Aviation Management, from Embry-Riddle Aeronautical University and was the first recipient of the Katharine Wright Award in 1981. She has received the "Woman of Accomplishment Award" from the Wings Club, The "Achievement Award" from The Aero Club of Washington and was inducted into the Women in Aviation "Pioneers Hall of Fame" in 1992. Her most recent tribute was bestowed during the 1992 National Business Aircraft Association meeting where she was honored by a group of her aviation peers for her many contributions to the continued support of general aviation.

Today Moya is a captivating speaker at national aviation events and college commencement ceremonies, where her inspirational message incites standing ovations and tearful applause. Her words of wisdom serve as proof, "You never know what you are capable of doing until you are challenged!"

Fran Grant & Jeanne McElhaton - Fear of Flying Clinic

Aerophobia Consultant, Fran Grant and Aviation Safety Analyst, Jeanne McElhaton teamed up back in 1976 to develop one of the aviation industry's most innovative enterprises, the Fear of Flying Clinic. Together they combined their aviation educational backgrounds and expanded existing aerospace programs to create a course that would assist people in overcoming their fear of flying.

Fig. 6-9 Fran Grant (l) Jeanne McElhaton (r)

Having spent most of her life in aviation, Fran first started flying in 1939 at Mills Field in San Francisco. She could barely afford 15 minutes of flying per week in a Luscombe but was able to complete her flight training with the help of a government scholarship in the Civilian Pilot Training Program. She joined the Ninety-Nines in 1941 and maintained an active level of participation in restoring the U.S. air marking system after World War II. She took part in many Powder Puff Derby starts until her health deteriorated, at which point she devoted her energy towards aerospace education. Fran became a senior educator with the Civil Air Patrol and an aviation education counselor with the FAA. She is also a member of both the World Aerospace Education organization and the American Society of Aerospace Education.

Jeanne McElhaton's multi-faceted aviation career includes more than 6,000 hours of flight time as a corporate, charter, commuter and ferry pilot, and more than 30 years competing as an air race pilot. She has taught all levels of pilot ground school instruction from private to instrument courses, flight simulator training and other related studies in aviation history and business. She acted as the host/instructor, writer and technical advisor for the TV series "Invitation to Fly" and has been a freelance aviation reporter for the CBS Radio Network since 1958. As an author, two of her works have been published; *VFR Flight Near TCAs: Pilot Practices, Perceptions and Problems* and *Terminal Control Areas: VFR Navigation in Complex Airspace.* Currently she specializes in analysis of report submissions from general aviation pilots and is a coordinator of an ongoing NASA/ASRS Human Factor workshop series. See Fig. 6-9.

After two crashes in World War II, Fran's husband developed a fear of flying which intensified as the years progressed. When his fear reached the point that avoiding airplanes and opting for alternative modes of transportation became the norm, Fran said,"We really have to do something about this." This was the catalyst that compelled the two aviation friends to combine their resources and create the popular Fear of Flying Clinics.

The fear of flying is an agonizing reality for more than 25 million Americans. For many the fear is innate and has developed from the most basic human instinct, the fear of falling. This problem is further compounded by those who feel that they are not in control while in flight, the discomfort of turbulence, claustrophobia and acrophobia. Even those somewhat comfortable with the flight itself are often confused about airports, ticketing, reservations and the entire operation that supports air travel. Both groups of people learn they are not alone and that relief is available through the clinics.

"We offer our clients the tools with which they can help themselves," explains co-founder Jeanne McElhaton. Based on the theory that "knowledge is power," they take an educational approach by bombarding the individuals with qualitative information about the aviation industry. The more the individuals learn about the system and how it works, the less fearful they will be about flying. This knowledge is combined with relaxation and control techniques which ultimately lead to behavioral change. Together these methods help to alleviate the very stress and anxiety that is connected with the flying experience.

Lectures are conducted by aviation professionals like; pilots, safety instructors, air traffic controllers, flight attendants and aircraft mechanics. Question and answer sessions examine the "Facts of Flying" and weather phenomena, and field trips tour air traffic control towers and provide hands on experience in airline evacuation drills.

Nearly all of the individuals "graduate" to the optional orientation flight at the conclusion of the Fear of Flying Clinics and continue to fly comfortably from that day on. Their horizons are literally broadened by those who had previously been bound to earth by their fears.

The clinic boasts a 90-95% success rate since Fran and Jeanne first helped to make the skies friendlier for the fearful back in 1976. Though the original clinic was slated to be a one-time event, the results were overwhelming and the concept quickly spread. Today Fear of Flying Clinics have been established on both U.S. coasts and in Australia. They continue to function with the cooperation of many individual volunteers from the Ninety-Nines, professional support from the FAA and major airlines like United and USAir and through the devotion of Fran and Jeanne, who still take part in the sessions.

Providing the worth-while service reaps endless rewards for the co-founders. "I'm overwhelmed at some of the things we have accomplished" exclaims Fran, "One just does not realize the tremendous amount of satisfaction there is in helping others. Due to my own physical infirmities, I actually receive more benefit assisting those that turn to me for help. I'm pleased to have done this, it has been a great joy."

A proud moment comes when the wings are presented to the graduates and Jeanne shares a poem to commemorate their new-found freedom of flight; "No longer bound to earth and all its' pains, no longer weighed down by chains of fear. Instead a look of joy, a sense of peace, a battle waged well fought and won. I celebrate for you dear friends, brave eagles that you are, each and everyone."

For their success in establishing aviation's leading service-orientated enterprise, the two women have received many honors. It is through their sensitivity and compassion that the thrill of flying has become a reality for thousands of land-lovers and their continued efforts in the unique program are to be commended.

Chapter Seven

New Frontiers

In the case of pilots, it is a little touch of madness that drives us to go beyond all known bounds. Any search into the unknown is an incomparable exploitation of oneself. Jacqueline Auriol

Geraldine Mock - Round-the-World Record Setter

We all know Amelia Earhart's brave attempt to fly around the world ended somewhere in the vast expanse of the Pacific Ocean. Over the decades since her death, there have been at least a dozen books on her and her achievements. There has been some wild and unfounded speculation on her ultimate fate and a group has even claimed to have found artifacts of her last flight. Conclusions have been drawn on circumstantial evidence but to date there is no conclusive proof that anything other than a navigational error caused her to miss the target and run out of gas.

The fact that no one attempted the feat for another 30 years after Amelia's death is a testimony to the dangers involved. But the world knew someday another woman would make the flight and conquer the goal. What the world did not expect was two women trying for the goal at the same time.

Geraldine "Jerrie" Mock and Joan Merriam Smith both flew around the world in 1964, as unintentional rivals, the first successful attempts at the feat by women since the disappearance of Amelia Earhart in 1937. Smith's route was a recreation of Earhart's famous, but ill-fated flight. Joan Merriam Smith, in an interview after her flight said, "I had the dream for years, first to fly an airplane, then to fly one as she [Earhart] did. When I was in high school, I would tell my friends and classmates that someday I was going to fly around the world just like Amelia Earhart. Everybody just laughed." [1] Smith, 27, married without children, earned her livelihood in aviation as a professional pilot. She had more than 8,200

Fig. 7-1 Jerrie Mock

hours flight time. Jerrie Mock, the 38-year old mother of three (ages 3, 16 and 17) was described at the time as a "housewife and amateur pilot." She had 750 hours and an instrument rating. (Mock was also an airport manager.) See Fig. 7-1.

For Jerrie Mock, the idea of being the first woman to fly around the world formed casually one evening in December, 1962. That fateful evening she mentioned to her husband, Russ, that, "I want to go somewhere and do something." His reply was in the form of a question. "Why don't you fly around the world?" She looked at him and he looked at her and they said, "Why not?"

Eighteen months later, after climbing mountains of paperwork both in the United States and every country where she would land or fly over, Jerrie Mock was almost ready to go. The Mocks had rounded up 22 sponsors, including some of the leading aviation suppliers. Smith, on the other hand, was sponsored by her hometown of Long Beach, California.

Smith's route would match Amelia's route, while Mock would fly a slightly different route. Jerrie Mock's route would involve several thousand more over-water miles than Smith's and one leg of Mock's route would take her over the Bay of Bengal, an extremely hazardous section of over-water flying.

Jerrie Mock's courage on selecting her route was understated at the time. She had only flown five cross-country flights and none of them were over water. She had taken her first flight lesson in September 1956, soloed nine and a half hours later and now seven and a half years later was attempting a round-the-world flight.

It should be remembered that Amelia Earhart had with her a navigator, Fred Noonan. Earhart had flown a 12-seat twin-engine Lockheed whose engines were rated at 450 horsepower. Mock planned on making the flight solo in a single-engine Cessna 180 land plane. Smith's plane was a twin-engine Piper Apache that weighed 5,000 pounds less than Earhart's and had a wing span 20 feet less. Joan Smith's Piper had engines rated at 160 horsepower.

Mock was asked why she was making the flight in a single-engine plane. Her reply was, "We just happen to have one." (The family owned the airplane.) [2] Mock chose the Cessna, she said, because it was more efficient than a light twin. It burned less fuel per mile, therefore she carried less fuel and less weight.

Both women would have the latest avionics including two Automatic Direction Finders (ADFs). The ADF, officials believe, would have saved Earhart.

Jerrie Mock was the first to file for the attempted record with the Federation Aeronautique International, and therefore had her flight sanctioned but Joan Smith was off the ground with a two-day head start over Mock on her 28-day trip.

Both women had their share of weather and mechanical problems but Mock also ran into some unusual preflight delays. Anonymous callers advised her not to make the flight and certain Congressmen were pushing a bill that would prohibit the U.S. Navy or Air Force from launching any search and rescue missions to help anyone "who was idiotic enough to undertake the adventure." There were unnecessary delays in equipment deliveries and the completion of modifications. Even on the morning of the flight some damage of unknown origin had to be fixed before the flight could begin. Throughout it all, Jerrie Mock remained cool and composed.

At 9:30 a.m. on March 17, 1964 Jerrie Mock's 11-year-old red and white Cessna 180 lifted off runway 10R at Columbus, Ohio. The practically new engine with only 65 hours lifted the plane off the ground effortlessly. In just two days, Jerrie and Russ Mock would be married 19 years.

Her flight, not unlike Amelia's, was fraught with danger, weather and mechanical difficulties. Inbound to Bermuda, Mock lost the radio beacon and then had to sit out a squall. She spent her 19th wedding anniversary on Bermuda as the house guest of Mr. and Mrs. John Fountain. He was also the FAI representative.

As a young girl, Jerrie Mock had watched movies of World War I dogfights and had slowly cultivated an interest in aviation. She also had a great-aunt who had been a friend of the Wright brothers. Her mother's maiden name was Wright and it frustrated her to know she was related to the aviation pioneer but did not have the documentation to prove it.

When asked why she was making the trip she said, "I'm doing it to give confidence to a little pilot who is being left behind in the jetstream of the space age."[3]

Later she said, "It was an opportunity to do something from which you really gain a great and deep sense of achievement."[4]

On April 11, 1964 Jerrie Mock landed safely back at Columbus, Ohio. She had covered 22,858.8 miles. Her time in the air was 158 hours, bringing her total time in her log book to just over 900 hours. Jerrie Mock had accomplished far more than being the first woman to successfully fly around the world. She set seven new records: a woman's speed record for a round-the-world flight; She set a speed record for men and women in a round-the-world flight in a single-engine plane in the 2,204-3,858 pound weight class; She was the first woman to fly the North Atlantic from the United States to Africa; the first woman to fly both oceans; the first to fly the Pacific in either direction in a single-engine plane; the first women to fly round-the-world as pilot in command; and the first to fly the Pacific west-to-east.

After several mechanical problems that caused delays, Joan Smith also completing her trip and landed safely. Surely Amelia Earhart would be smiling, for her spirit flew with both Joan Merriam Smith and Jerrie Mock.

First Lady Astronaut Trainees

In the late 1950s, America embarked on its journey into the last frontier - space. Predictably, the question came up again. Can women experience space flight with no adverse reactions? There was little data at this time to judge how men or women would react in space. Experiments seemed to indicate that men could handle the rigors of space flight if properly selected and provided with the appropriate equipment. Only a few women had flown jets and approximately eight women had flown through the sound barrier, but women had not had the opportunity to become jet test pilots or participate in many of the experiments to determine human physical adaptation to space. However, the WASPs had scored high marks in the hyperbaric chamber during World War II. There still seemed to be a profound ignorance in NASA about women's capabilities to adapt to space flight. Little was known at the time about human space physiology, since only a few humans had actually gone beyond the stratosphere. Somehow the mysteries of space and the inherent dangers of a airless environment caused NASA to draw conclusions that women were not the appropriate persons to face the dangers of space flight. There were, however, a few who stressed the scientific approach. They said "Let's test a woman and see."

In September 1959, the Air Force began highly confidential experiments on the physiological or psychological characteristics and suitability of women for space flight. Geraldine "Jerrie" Cobb, a young and skilled professional pilot, met Dr. W. Randolph Lovelace of the Lovelace Foundation, in Albuquerque, New Mexico. Also there was Brigadier General Donald D. Flickinger of the Air Force, and both men were distinguished in the field of aerospace medicine. At the time, Cobb, 29, had been flying since she was 12 years old. She was licensed at 16 and flying commercially at 18. In accumulating an outstanding record, she had logged 10,000 hours of flight time, (including jets) set three world records, and earned the Federation Aeronautique Internationale gold wings of achievement. She was a pilot and manager for Aero Design and Engineering Company, which manufactured the Aero Commander aircraft. She was one of the few women executives in aviation at the time. Cobb gave up this career to become the first woman to undergo the Mercury astronaut training.

Geraldine Cobb completed and passed a battery of 75 physical tests in February, 1960. Many of the tests resembled a medieval torture. The eye exam alone took four hours. They tested her nerves with electrical impulses. She was strapped on a table, tilted to a 65 degree angle, tested, tilted and tested again and again. Every five minutes she was given an electrocardiogram. They taped wires to her body and measured her oxygen intake while she peddled a stationary bicycle paced to a metronome: then they kept adding 15 pound weights to the point of exhaustion. They injected ice water into her ears to induce vertigo and measured her recovery time. Teams of experts then attempted to anger her with questions designed to measure her evenness of temper. Not a single inch or function within the person of Jerrie Cobb escaped analysis. The results came forth as one word: EXCEPTIONAL!

In the summer of 1960, Cobb successfully passed the second phase of testing, the psychopsychiatric tests. These included a nine hour and forty minute record stay in "profound sensory isolation." She remained in an eight foot tank of water that tests the subject's mental resources during depravation of the five basic senses of sight, hearing, taste, touch and smell, while in a simulated weightless environment. One male, after four-and-a-half hours in the isolation tank, began babbling, hearing dogs bark, singing obscene songs and broke into tears at the state of humanity.

In April 1961, Cobb underwent a two-week series of stress tests at the U.S. Naval School of Aviation Medicine at Pensacola, Florida. This was the third phase of testing her physical and mental capabilities to endure space flight. She satisfactorily performed all the standard tests given to Navy pilots and several experimental tests for space flight.

Cobb successfully passed all the tests used to select the original seven Mercury astronauts. She was qualified for space flight! In May 1961, the same month Alan Shepard became the first American to fly into space, NASA named Jerrie Cobb a consultant. She later commented, "I was the most unconsulted consultant in NASA." The House committee that later investigated the subject of women in space heard from one of its committee members on the issue. "She held a position and says what is to be done, but is not allowed to do it."[5] Cobb, in the meantime, watched as one male candidate after another was selected for the space program.

NASA assumed that Cobb's results must have been an aberration and asked Dr. Lovelace to see if the exceptional results from Cobb's tests could be repeated by other women. This decision would later come back to haunt them.

With the help of Jacqueline Cochran, twenty-five women, all of whom were outstanding aviators, were selected for the tests. It was hard for a woman pilot to find a job in the man's field of aviation, but these women quit well-paying jobs to take part in the tests. That is how serious and committed to the space program

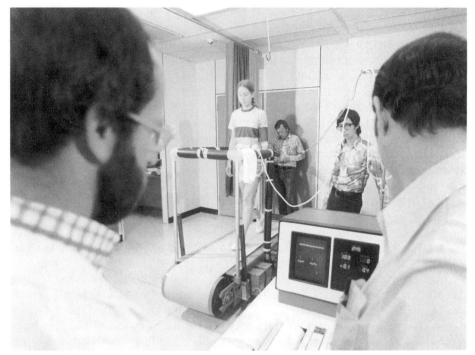

Fig. 7-2 NASA engineer Marsha Ivers sets baselines in 1976

they were. For those women who could not afford the expenses, Jacqueline Cochran provided assistance from her personal funds. There was no taxpayers money used for this experiment.

The tests were so comprehensive that scientists could tell accurately what part of the country each woman was from by the level of Strontium-90 in her tissues. In all, the women had 87 different tests, 83 X-rays, reaction tests, shock tests, 18 needles stuck in their heads to record brainwaves, and an assortment of other tests.

Of the twenty-five candidates, twelve successfully passed all the tests. NASA had a new problem. The test results turned out too well. NASA had Cobb and twelve other women eager to fly into space, but NASA had never planned on successful results or on putting women into space.

The Pilot Candidates:

Jerrie Cobb - CA	Jane Hart - MI
Bernice Trimble (Steadman) - MI	Jean F. Hixon - OH
Myrtle T. Cagle - GA	Rhea Hurrle (Woltman) - TX
Jan Dietrich - CA	Irene Leverton - CA
Marion Dietrich - CA	Jerry Sloan (Truhill) - TX
Mary "Wally" Funk II - OK	Gene Nora Strumbough (Jensen) - TX
Sarah Lee Gorelick (Ratley) - MO	

Fig. 7-3 14-years later astronaut Marsha Ivers in space

NASA solved the problem by refusing to authorize the completion of the next phase of tests. They felt that allowing the tests to continue might be taken as "approval" of female astronauts. (NASA also feared a backlash to the entire program if a woman were to die on a mission.) The decision to cancel the tests was made just one day before the women were due to leave for Pensacola. There was bitter disappointment and anger from the women toward the NASA administrators. The resulting controversy involving NASA, Congress, and a handful of determined young women set the stage for a decade of furious debate on the role of women in aviation.

The tests of the women had been conducted in secret with one or two women at a time. The women had been asked not to discuss the tests with anyone, alleging that if they were made public there might be many questions and even cancellation of the tests. The women agreed and did not discuss their situation with anyone. Many never knew who or how many other women were in the program. In fact, several did not meet until 30 years later at a reunion at the International Women's Air and Space Museum, in Centerville, Ohio.

Cobb was outraged by the cancellation of the tests and went searching for answers. Jacqueline Cochran, who supported the test program wrote a personal letter to Jerrie Cobb: *Women for one reason or another have always come into each phase of aviation a little behind their brothers. They should, I believe, accept this delay and not get into the hair of the public authority about it. Their time will come and pushing too hard just now could possibly retard rather than speed that date.*[6]

Cobb did not agree with Cochran. When she and Jane Hart, wife of Michigan Senator Philip Hart, and one of the successful test candidates herself, could not get a satisfactory answer to why the tests were canceled, they took off the self-imposed gags and insisted on their right to participate. They lobbied Congress, Vice President Lyndon Johnson, and NASA for inclusion of women in the astronaut program. Until Allen Shepard had flown into space, NASA steadfastly disclaimed any program for women in space. When the press finally caught on to the testing program they typically, skeptically and flippantly tagged the women with nicknames which none of the women found endearing.

Cobb and Hart managed to get the House of Representatives to hold a two-day hearing on the issue. In the Congressional hearings both Jerrie Cobb and Jane Hart testified. Cobb introduced a litany of excellent test results, some of which were better than the men's results. She provided strong medical and economically sound reasons to include women in the space program. She testified, "Women weigh less, consume less food and oxygen than men, a very important point when every pound of humanity and the necessary life support systems is a grave obstacle in the cost and capability factors of manned space vehicles. . . . Women are less prone to heart attacks... scientists say women are less susceptible to monotony, loneliness, heat, cold, pain and noise..."[7]

Both Cobb and Hart faced baited questions; did they think women were better than men? Hart cut to the heart of the issue. "It is inconceivable to me that the world of outer space should be restricted to men only, like some sort of stag club."

Hart was not asking for preferential treatment. "I am not arguing that women be admitted to space merely so that they won't feel discriminated against. I am arguing that they be admitted because they have a very real contribution to make."[8]

When NASA realized that women were as qualified for space flight as the men, they immediately changed the requirements. NASA announced a new formal requirement that a candidate must have experience as a jet test pilot. Prior to this time all of the male candidates selected had coincidentally been experimental jet test pilots, but it was not a stated requirement. This new requirement precluded the chances of selection for every woman in the United States. (Prior to this change, the criteria were a U.S. citizen, have a commercial certificate with an instrument rating, preferably an ATP (ATR in those days), a college education, and be between the ages of 25 and 40.

John Glenn and Scott Carpenter, both astronauts, presented the "male point-of-view." Glenn said that the requirement that an astronaut be an engineering jet test pilot was fair. There were decisions and judgements necessary in space flight and that jet test pilots had the experience to make these decisions. He insisted that he was "... not anti any particular group. I am just pro space."[9]

Cobb countered that there was a factor called the equivalent experience in flying that may be even more important in piloting spacecraft. She said that, "Pilots with thousands of hours of flying time would not have lived as long without coping with emergencies calling for microsecond reactions.

"What counts is flawless judgement, fast reaction, and the ability to transmit that to the proper control of the aircraft. We would not have flown all these years accumulating thousands and thousands of hours in all types of aircraft without accumulating this experience."[10] (Collectively, excluding Cobb's 10,000 hours, the women averaged more than 4,000 hours compared to the men's 2,000 hour average. None of the male astronauts could come close to matching Cobb's record. John Glenn and Scott Carpenter trailed far behind with 5,100 and 2,900 hours respectively.)[11]

A committee member pointed out that Glenn and Carpenter did not have the required engineering degree when they were selected for the program. Glenn responded by saying, "Mine was the equivalency thereof, and it was felt, with my in-service experience and schools I had been to, while I did not have the actual hours at college, I had more than the equivalency of an engineering degree."[12] Glenn thought the equivalency factor was appropriate for him but later ruled it out relative to Cobb's 10,000 hours as appropriate in lieu of test pilot hours. Apparently having the"right stuff" also meant being the right sex.

Cobb pushed for a compromise by suggesting that pilots without test pilot experience (the women) go through the simulator see how they do, and compare the results to the men.

Cobb went on the record saying that the Navy would not allow her access to the simulator to demonstrate her capability for space flight. "I find it a little ridiculous," she said, "when I read in a newspaper that there is a place called Chimp College in New Mexico where they are training 50 chimpanzees for space flight, one a female named Glenda. I think it would be at least as important to let the women undergo this training for space flight."[13]

James Fulton, Representative from Pennsylvania, supported, at least in words, the women's quest and added to Cobb's comment: "When the scientists first started putting living animals in space, it is rather remarkable that both the dogs and monkeys were all without exception female.

"Both Russian and American scientists put female animals into space and suddenly they stopped, when they found out they were successful. I think this is very remarkable."[14]

A NASA official testified to the committee that training women would, "be a waste, a luxury the nation could not afford if they were to get a man to the moon by 1970."

Jacqueline Cochran, who had been invited to testify on two day's notice was once again in the middle of the argument. Although Cochran paid many of the test program expenses for some of the women, she testified during the special

Congressional hearings against the idea of including women in the space program. She believed that the only way to include women was through a trained corps of specialists, of at least 100-150 (a concept strongly reminiscent of the WASPs). Cochran said the nation could not afford the time or money to conduct such a program. She spoke of women's lack of commitment and high attrition rates due to marriage and pregnancy. She reminded the committee that according to her figures the attrition rate among the WASPs was 40 percent.

There is strong speculation that Cochran did not actively support the effort to put any of the 13 qualified women into the space program because she was beyond the age limit and could not participate in the program herself.

The two-day hearings of the House of Representatives Special Subcommittee on the Selection of Astronauts were shot down by many Senators, Congressmen and attending male astronauts. Two excuses were put forth why women would not be permitted to train for space flight: no funds and no engineering degrees. In hindsight, the reader may draw other conclusions.

In June 1963, the Russians launched the first woman into space, twenty-six-year-old Valentine Tereshkova. It was political one-upmanship. Tereshkova was not a jet test pilot or an engineer, but a textile worker and her aviation background was that of an amateur skydiver. Tereshkova made 48 orbits in the *Vostok IV.* Her presence added prestige to the flight and she did go on to become a respected international spokesperson for the Soviet Women's Committee and for Soviet women.

The Russian's agenda was similar to the United States. They did not have time to train women for space, and it would be twenty years before they sent another woman up in space, and then it was to upstage Sally Ride's launch.

In 1964, NASA announced a new policy to recruit an elite corps of scientist-astronauts for the upcoming Apollo and Skylab missions, and they encouraged women to apply. Perhaps this was the breakthrough? Many qualified women applied, but not one was chosen, at least not for the next fourteen years.

During those barren years women were tested and base lines for performance were developed. They were no different than those set by the original 13 women astronauts.

Science fiction writer Robert Heinlein echoed what many women in the United States felt: "If you want my personal opinion, women haven't been invited into the space program because the people who set up the rules are prejudiced. I would like to see some qualified women hit NASA under the Civil Rights Act of 1964. They're entitled to go; they're paying half the taxes; it's just as much their program as ours. How in God's name NASA could fail to notice that half the human race is female, I don't know."[15] See Fig. 7-3.

The women who participated in these experiments were human guinea pigs, but their efforts were not in vain. The excellent results served as the base line for later women like Eileen Collins and Kathryn Thornton who did reach into the final frontier.

Kathryn Thornton - Astronaut

Kathryn Thornton had charted a course for space when she graduated from Auburn University in 1974, with a B.S. in physics. From there she earned an M.S. and Ph.D in physics from the University of Virginia.

Before joining NASA, Dr. Thornton worked as a physicist for the U.S. Army Foreign Science Technical Center. She joined NASA in 1985.

On May 14, 1992, Thornton took her second space walk as she and Tom Akers stepped from the relatively safe protection of the Shuttle *Endeavour* to the cold and vacuumed environment of space. Their job was to practice space construction techniques by building a tinkertoy-like pyramid with aluminum tubes and joints. *Endeavour*'s first flight, cited by NASA as a symbolic recovery from the 1986 explosion of *Challenger*, quickly captured the public's attention like no other recent mission. Even NASA's space shuttle director, Leonard Nicholson, was excited. "To call what we have all seen in the last few days exciting is a gross understatement."[16]

Thornton and Akers' walk had been delayed because of difficulties with the primary mission of rescuing a stranded satellite and reinserting it into its proper orbit. *Endeavour* Commander Dan Bradenstein originally planned to use

Fig. 7-4 Astronaut Kathryn Thornton

Thornton as one of the two astronauts to rescue the stranded satellite. However, Akers, who outweighs Thornton, 170 pounds to 115 pounds, was chosen instead by Bradenstein when the team decided that Pierre Thuot and Rick Hieb, the other astronauts aboard, should "manhandle" the 8,900 pound spacecraft into the cargo bay. Thornton and Akers did perform a seven hour space walk and tested several devices including some that could help "overboard aeronauts" get back to the shuttle safely. See Fig. 7-4.

The only extraordinary coverage Thornton received was the notice in at least one newspaper that she was the "first mother to walk in space." (She has three daughters and two stepsons.)[17]

Thornton has over 333 hours in space and more than seven hours space walking. She is also one of 21 women currently qualified as astronauts.

Major Eileen Collins - Astronaut/Shuttle Pilot

One reason Major Eileen Collins became the first woman Space Shuttle pilot is her heavy background in mathematics and science. Traditionally, she says women have not been encouraged to major in these subjects and feels that when more do, there will be more opportunities opened to them.

Major Eileen Collins was born in Elmira, New York. She earned her first degree in mathematics and science from Corning Community College in 1976, and her second, a B.A. in mathematics and economics from Syracuse University in 1978. Major Collins then went on to earn a Masters of Science in operations research from Stanford University, and another Master of Arts in space systems management from Webster University. See Fig. 7-5.

Major Collins graduated from Air Force Undergraduate Pilot Training in 1979 and became a T-38 instructor. She held that position until 1983 when she became a C-141 aircraft commander and instructor pilot at the Air Force Academy. Major Collins went into the astronaut program while attending Air Force Test Pilot School at Edwards Air Force Base. She graduated from that program in 1990, and in 1991 qualified as a pilot on a future Space Shuttle flight crew.

Collins recalls that in 1978, when she graduated from college, there were already six women in the space program. She thought to herself, "Hey, I can do this," although this was not her goal as a child. She is now thankful she had role models and tries to be one herself.

The tests she passed to become an astronaut were no different from those the men take. Of the 186 candidates interviewed and tested, 23 passed. Collins is among the five women selected.

One of the toughest tests, learning what zero gravity feels like, Collins found to be "the most fun I ever had in my life." Nicknamed the "vomit comet" by some of the more queazy personalities, the KC-135 took Collins and her classmates on a roller coaster ride from normal gravity to zero gravity and on to 2 "Gs" of gravity more than 40 times a day, on and off for several weeks. "You get used to it," she says. "You build up a tolerance."

Fig. 7-5 Major Eileen Collins

Collins practiced Shuttle landings in a highly modified Gulfstream II and is currently assigned to Orbiter Systems, modifying the Orbiter electrical, environment, power and propulsion systems. In her off-duty hours she enjoys running, hiking, golf, reading, photography and astronomy. Her burning desire these days is to strap on the Shuttle, break away from the hold of earth on top of a rocket with the equivalent thrust of 200 DC-9s and see her planet earth from space.

Marta Bohn-Meyer - Test Pilot - SR-71 *Blackbird*
On October 3, 1991 Marta Bohn-Meyer became the first woman in history to fly as a crew member aboard the triple sonic SR-71 aircraft. See Fig. 7-6

Marta and her husband, Bob, are both flight test engineers assigned to the SR-71 program at NASA's Dryden Flight Research Facility. They have been involved with the high speed project from its inception. "When the Air Force decommissioned the airplanes back in 1989, we were assigned to examine the logistics of obtaining and implementing the aircraft into the NASA programs."

The Blackhawk SR-71 is now being used for high speed and high altitude research that can be applied to the design of future civil and military aircraft.

Extensive training is required for a mission that will take you three times the speed of sound. As project manager of the high speed project, Marta spent more than 150 hours in the SR-71 simulator just to prepare for her first flight. That first flight was not only the most thrilling event for Marta but also the most draining and exhausting. "There are so many duties and tasks to perform in the airplane that you really don't have time to sit back and contemplate your surroundings. I did look up to see the Great Salt Lake whiz by but for the rest of the flight I had other things to concentrate on."

Marta describes her historic flight as out of this world. "Flying at Mach 3 is at the top of my list of experiences and there is nothing that could equal it for me. I had a grin on my face from ear to ear and we hadn't even left the pavement yet. If I ever get down or depressed, I just remember how great that feeling was and it gets me motivated again."

Marta was born August 18, 1957 in Amityville, NY. "My parents wanted to help me choose a hobby when I was growing up and I narrowed it down to horses or airplanes. My father worked for Grumman, a major aircraft manufacturer, and I decided I knew more about airplanes than I did about horses, so for Christmas my parents gave me introductory flying lessons." Marta was flying at the age of 14, and earned her private pilot's license weeks before her drivers license.

"Many of my dad's friends were test pilots and I was so enamored with them. When I had to do reports or stories in school, I'd pick these people to interview because I thought that what they did was so interesting. Ultimately I knew I wanted to be a test pilot but young girls just didn't go around saying that's what they wanted to be."

Marta knew the chances of reaching this goal would be slim because most test pilots have military backgrounds and the military wasn't accepting any women in flight school. Marta attended Rensselaer Polytechnic Institute, in Troy, N.Y and studied aeronautical engineering. Polytechnic and NASA had a cooperative education program at Langley Research Center in Hampton, VA and Marta became a part of that program. She participated in rotorcraft research, wind tunnel and flight safety projects associated with civil aircraft.

She graduated in 1979 with a Bachelor of Science degree in aeronautical engineering and NASA's Dryden Flight Research Facility hired her as an operations engineer. She worked on a variety of research projects, including flight tests on the Space Shuttle's thermal protection tiles.

The prospect of entering a profession dominated by men didn't intimidate Marta. She was used to flying with male pilots as equals. It wasn't until she was a senior in college that she discovered people might treat men and women differently when it came to employment opportunities.[18]

Fig. 7-6 Marta Bohm-Meyer

Marta approached a NASA test pilot who was running the flight operations division and told him that she had come out to Dryden because she hoped to get an opportunity to fly. He gave her his full support.

In breaking new ground, Marta attributes her success to being in the right place at the right time, and having many male supporters who opened doors for her. "All the people who supported me in my career goals have been men. There really haven't been any women who could positively reinforce my career path because there weren't any women in the positions."

Being part of a triple sonic team is just the beginning for Marta and her husband. As pilots they fly in aerobatic competitions using airplanes that they build in their spare time. Together they have built three Pitts Specials. "We were building the airplanes in our garage until the operation got so big that we had to take time out to build a permanent shop." As true engineers they have made modifications to the original aircraft specifications to increase the plane's performance and maneuverability. "Some of our experiments have not been worth the energy but our kevlar spinner for example has turned out to be a pioneering design that has caught on in the aerobatic world."

The couple compete in different categories of aerobatic competition, Marta in the Intermediate category and Bob in the Unlimited. "We are very supportive of one another and don't compete against each other. Our approach to flying is also very different, I'm more of a seat-of-the-pants pilot and tend to look at flying more passionately, while Bob's approach is analytical." In 1990, both won California Championship titles in their respective categories. That was a very rewarding honor for the two to share.

People assume Marta's next goal is space flight. "Many people are disappointed when they hear that I'm not interested in becoming an astronaut. I'm totally contented with the challenges and responsibilities of my position at NASA's Dryden Facility. From an engineering standpoint, it's very fulfilling to be involved in every level of a project like the SR-71." Whether designing the programs, coordinating the missions, or maintaining the flight proficiency of the crews and the aircraft, Marta obviously has the best of both worlds.[19]

Rose Loper - Test Pilot - Boeing Airplane Company

Rose Loper was born in 1948 in Eustis, Florida, and came of age at a time when women traditionally became teachers, nurses and secretaries. Opting for teaching, Rose attended Carson-Newman College in Jefferson City, Tennessee, and graduated with a BS in physical education. She went on to complete graduate school, earning her MS in recreation administration from Florida State University in 1973. "I found a teaching job right away in Richmond, Virginia, but it didn't take long for me to realize that I had made a mistake and that teaching was not for me. What a dilemma to have spent all the time, money and energy on an occupation that I really didn't enjoy."

Rose was intrigued by courses she had taken in graduate school that examined military recreation programs and their efficiency and organization. That interest, coupled with family members who had been in the military, helped Rose decide to join the Army. She was offered a direct commission as a Second Lieutenant and was detailed to the Transportation Corps, focusing primarily on helicopters at Ft. McClellan, AL.

Rose was interested in flight school, but in 1974, women were just beginning to break into flight training programs in the Army and the Navy and the routing was not well defined. One day in class another officer questioned the instructor about how to get accepted into helicopter flight school and was told that you simply take a test and if you pass you can go. A show of hands indicated that many of the officers were interested, including Rose. Twenty-five of Rose's fellow women officers enrolled to take the test but only 12 actually arrived on the assigned testing date. Six of the women passed but two were later medically disqualified, two went to Europe and one gave it up to get married.

"Getting accepted into Army flight school wasn't as easy as the instructor had implied, I found that it didn't involve just passing the test. There were additional phases that I had to apply for, different departments to contact to express interest, physicals and orientation flights to undergo, interviews to pass etc. I persisted, acknowledging each requirement that was put before me as a

Fig. 7-7 Rose Loper

challenge. I came to realize that the doors were open for me if I was just smart enough to figure out a way to step through." See Fig. 7-7.

Personal perseverance paid off and Rose reported to Army Flight school in Ft. Rucker, AL in the fall of 1974 as the 14th woman accepted into flight school. "The first woman was in the process of graduation as I arrived. I had orders to report to a specific building at a certain time and I showed up to find I was the only woman. To my classmates' surprise, I answered when the roll was called, and I later learned that they all thought I had arrived to take the roll!" Rose fell in love with flying immediately and had a natural touch that made her enjoy her training period. In May of 1975, Rose graduated as the first woman to graduate with honors.

Focusing on aviation maintenance next, Rose was the fourth woman to go through the Aviation Maintenance Officer Corps Course and upon graduation she became a test pilot. "Maintenance test pilots are the individuals who are dumb enough to fly equipment after repairs. I always insisted that the lead mechanic accompany me on the test flights as an added incentive to do the job right the first time." Rose eventually became the production control officer and managed aviation maintenance and direct support for the 9th infantry division. She was promoted to Captain in 1978 and took command of a 280-person ground maintenance company at Ft. Lewis, Washington.

In Rose's spare time she used her veterans' benefits to get her fixed-wing license, including her commercial and instrument ratings. Rose fell in love with the Northwest while serving at Ft. Lewis and left the Army in 1980 to work for Boeing in Seattle, WA. With her maintenance and flying background, she found herself right at home as a ground operations engineer on the Boeing 767 Aircraft Certification and Flight Testing Program. Rose was the first woman ever hired by Boeing for that particular job. She worked simultaneously with the engineers who design and develop the aircraft as well as the mechanics who configure it for testing. "It was a highly visible position and required a lot of coordinating. I received information about the testing required on an aircraft and gave instructions to manufacturing on how to configure the airplane for actual testing. I was physically onboard the airplane from the time the first rivets were drilled into the sheet metal up through the final flight certification phase. From this vantage point I learned that the pilots had the best job and let the fact be known that if they ever needed a pilot I was available."

In April of 1983, Rose got her chance to fly when she was offered the position as Boeing's Company Corporate Helicopter Pilot. "What a challenge, to be the first woman pilot at the Boeing company. I had only flown in the Army and then as a part of a pilot/co-pilot team, now suddenly I was the Boeing Company's corporate helicopter pilot. I worked hard to pass the ground school for the aircraft and quickly went to work flying Boeing executives as well as major airline executives and heads of state from all over the world."

Rose was very conscientious and safety orientated and did everything possible to represent her company in the best light and to make her flights the best experience for all her passengers. "One day I flew over to the Boeing Commercial Aircraft Headquarters building to pick up some Asian customers. I parked out on the front lawn and went in to greet them. They were courteous but evidently thought that I was their guide. After I got them all buckled in, I climbed in the pilot's seat and prepared to fly away. The startled Asian executives looked at me wide-eyed and said: 'You qualified? You certified?' With my engines running I lifted off and said, 'You betcha.'"

While I thoroughly enjoyed my corporate position, my friend and confidant who had helped open many doors for me at Boeing, reminded me that Boeing didn't build helicopters in Seattle. This turned out to be the best piece of advice I would ever receive. I took that to heart and started building my fixed wing experience flying, as a co-pilot on some of the test fights. The biggest transition for me was that of speed, having spent my life at 90 knots in helicopters compared to 390 knots in the big jets. It was a big learning curve to overcome but I kept on trying."

When Boeing re-organized in 1986, Rose was assigned as the chief pilot in the flight test department where she started working in the production phase. This not only improved Rose's opportunities to fly but prepared her for Boeing 737 flight training and her ATP rating. By 1989, she was a captain on the Boeing B-737 and in 1990 a captain on the B-757. In 1990 Boeing re-organized again and this time Rose was promoted to a full-time position as a production test pilot. Of all the pilots at Boeing, thirty-five are production test pilots, thirty-three are men and two are women.

As a production test pilot, Rose is part of a crew of five who fly the jetliners as they roll off the assembly line to perform airborne system checks. "It's a magic moment when a shiny new Boeing jetliner leaps into the sky for the first time using maximum thrust at minimum weight. Onboard we isolate every system down to the lowest failure mode of continued operation and every engine is shut down and re-lit using each igniter."

One of the highlights of Rose's career has been her involvement in the Army's SDI program (Strategic Defense Initiative.) After learning that Boeing would be doing the testing, Rose lobbied with other Boeing pilots involved in the project to be assigned to testing. "I got the job because of my background but also because of my persistence." Rose was in the co-pilots seat for the first fully-equipped flight on May 11,1990. This multi-million dollar experiment required the flight crewmembers to do tracking, timing, turning, reprogramming and testing from 46,000 feet. The onboard sensor is sensitive enough to pinpoint the heat-emitting objects in space and can detect the heat of a human body at a distance of more than 1,000 miles. "As an early detection system, the earlier you can detect an aggressor, the better. In conjunction with the entire SDI program, it is another addition to world peace."[20]

Rose continues to fly in the U.S. Army Reserves and is currently Company Commander of A Company, 6th Combat Aviation Battalion, 158th Regiment. Not limiting her interests to just land and air, Rose is an avid sailor and plans to fulfill her dream of sailing around the world when she retires from Boeing.[21]

Susan Darcy - Test Pilot - Boeing Airplane Company

Susan Darcy was hired by Boeing as an engineer's assistant in 1974 when opportunities for women in the aviation industry were just starting to take off. Susan always knew that she wanted to fly and in 1977 she started taking flying lessons. She progressed quickly, earning all of her ratings and completing her college education in 1981 with a B.S. in Aeronautics and Astronautics from the University of Washington in Seattle.

From 1982 to 1985 Susan worked in Boeing's 757/767 training department, teaching airline pilots the latest in Boeing technology. The new state of the art equipment required systems analysis and cockpit procedures which Susan helped to develop. She wrote and produced much of the aircraft's training material including the flight controls, hydraulics and landing gear chapters for the B-757/767 operations manual. See Fig. 7-8.

Susan continued building her flight time and experience after hours in hopes of furthering her aspirations to fly. "I was accustomed to setting goals for myself and felt a need to focus in more specifically on my target. I was fortunate to have grown up in an environment that encouraged me to believe I could be anything I wanted to be, regardless of gender. With that in mind, I imposed on myself a

Fig. 7-8 Susan Darcy w/co-pilot Leon Robert

one year 'drop dead date' of November 1, 1985 for my goal of landing a professional flying job and pursued it feverishly. On October 31, 1985 at 12:30 p.m. with just 11 1/2 hours to go, I was offered the job as Boeing's first female production test pilot."

Susan's position as a Boeing test pilot requires her to be rated in the full line of Boeing equipment. She received her B-737 type rating in 1987 and her B-757/767 type in 1989. She was the first woman to be rated as a captain on the new B-747-400 when she earned that airplane's type rating in May of 1989. In 1991, Susan became the first woman rated as an instructor for the Boeing Company on the B-757's and in 1992 she completed her training to instruct on the B-737's.

Besides flying all the makes and models of airplanes that Boeing builds, the production test pilots must also be proficient in both the left and the right seats. "A typical week might have me scheduled on Monday as captain on the maiden flight of a new B-737 and on Tuesday in the right seat of a B-757 with a customer pilot checking out his new airplane. On Wednesday I might be on a flight to say, Las Vegas, for an aircraft sales delivery, and on Thursday, doing a maintenance follow-up flight on a B-747-400 etc. And when the airlines are too busy to take delivery of their own airplanes, we travel abroad to places like China, Korea, or Australia to name a few. This diverse scheduling, coupled with varied equipment, seat positions, destinations and job descriptions is very challenging to me."

"I feel very lucky to be able to combine my life long love of flying with a position at Boeing that is very rewarding and to get paid to do it all. It's a daily affirmation for me that goals that I set became a reality."

Jeana Yeager - Pilot *Voyager*

Jeana Yeager grew up with parents who never assigned gender-related limitations to her. As a result, she had options like dressing up and being "lady-like" or being a "tomboy" and climbing trees or hunting tadpoles. She also grew up around horses and they still influence her life. Her first horse, (actually her older sister's horse) named "Buffalo" would become the foundation for her lifelong philosophy of persistence to accomplish a task.

Jeana was only three or four when Buffalo arrived and she was so small she could walk under the horse without bending over. At first she just walked the horse but soon Yeager wanted to ride him. She was too small to get a bridle and saddle on Buffalo but she knew she could find a way. Her resourcefulness led to a fence. When that failed, she found a tree stump, and when Buffalo still resisted carrying her as a passenger, Jeana walked him to a picnic table. Each time she would try and climb aboard the horse, Buffalo would back away. She persisted. Jeana moved the table over to a tree and wedged the horse in between. It worked. But her next problem was getting Buffalo to go where she wanted him to go. Without a bridle she solved that problem by finding a tall bamboo pole. Now all she did was tap the pole on the side opposite to the direction she wanted to go.

In high school, Jeana ran track and that taught her teamwork, and what trying, succeeding, winning and losing are all about.

Even though Jeana Yeager did not grow up in aviation she remembers an early fascination with helicopters. As a child, helicopters reminded her of dragonflies, hovering and maneuvering in the air. At the age of 26, Jeana decided it was time to do something about the constant, nagging desire to fly. She went down to a company in Santa Rosa, California that operated a flight school. The people running the flight school told Yeager that she was better off getting her fixed-wing rating first, and then her helicopter rating. Yeager went ahead and earned her fixed-wing rating in 1978, but the company fell on hard times before she could begin her rotorcraft training.

Yeager had experience in various types of drafting: illustrating, mechanical, geophysical, geological, and some architectural. She had worked for a company involved in off-shore drilling, where she did seismic mapping. Later she went to work for a company called Project Private Enterprise, a privately funded program to put a person into space. This turned out to be a rewarding experience for Yeager. She worked for a fellow named Bob Truax, who answered all her questions and gave her a free hand in wandering from the office into the shop area. Captain Robert Truax, USN (Ret). had been experimenting with rockets since 1937, and is a true rocket pioneer. He is one of the few people who worked with both Robert Goddard and Werner Von Braun. He worked on the rockets that went into the X-1 experimental airplane, helped create Jet Assisted Take Off (JATO) and launched rockets from the ocean. Truax was forming his own company to launch rockets into space when Yeager met him.

"Bob tried to keep me in the office with aeronautical drafting, but I kept drifting off into the shop." Yeager ended up doing office work as well and became the person who did a bit of shop work, too. "It was a wonderful friendship with him and all the people who were there - a very good all-around work experience. I learned a lot of good basics that served me for the *Voyager* program."

Being in the right place at the right time some have said is the key to success. Yeager doesn't quite say it that way. "Time and circumstances got me into flying experimental aircraft. I met Dick Rutan in 1980, and we started friendly competition flying and setting records. I hold five world records (in speed and distance), and Dick has six world's records. Then we went on to the *Voyager*, which seemed a nice evolution to the next records.

At lunch one day, the talk got around to a round-the-world flight, on one tank of gas. While the two brothers discussed the various records, Jeana said: "Why not? Let's do it." The two brothers could not think of a reason not to.

"Once we decided to do the *Voyager*, we more or less rolled up our sleeves, and not knowing a lot about anything, set up a corporation...learning how to operate a corporation, putting together an airplane without plans, making it up as we went, figuring out how to raise funds, and how to make things happen."

Yeager, the 34-year old co-pilot of the round-the-world, non-stop flight of the *Voyager*, and the pilot and designer of the aircraft, Dick Rutan (Burt Rutan, Dick's brother, was co-designer) considered the flight, "the last big plum in aviation history." Rutan accepted Yeager with unqualified conviction. By now Yeager held several world flying records, some of which Rutan himself had once held.

There was extensive media coverage of their project.Very aggressive marketing was also necessary to raise more than $2 million needed for the flight. Yeager, who likes a low profile, left the public relations and the actual radio contacts during the flight to Rutan, as much as possible. He said, "She likes to get on with the job, and she is very much an equal partner in this."

Throughout the preflight marketing of *Voyager* there was no emphasis on one of the pilots being a woman, and Yeager did not receive any unusual or individual attention.

On December 14, 1986, after six years of planning, building and testing, the *Voyager* took off on its round-the-world flight. The flight went without any serious difficulties or setbacks. The physical and emotional strain the pilots endured, however, was extraordinary. The aircraft had a wing span greater than a Boeing 727, and weighed only 2,000 pounds. It carried four times its weight in fuel.

Nine days of living in a cabin seven feet long, and four feet high, tested to the limit the private relationship the two had shared for more than five years. Comfortable sleep was almost impossible since the very light aircraft reacted

Fig. 7-9 Jeana Yeager

violently to any turbulence. In fact the most sleep either pilot got was about 3 to 4 hours of catnaps. The constant engine noise threatened permanent hearing loss, and increased the psychological strain. Each time they shifted positions they had to rebalance the fuel to load maintain stable flight.

A primitive, if not pioneering, environment existed aboard *Voyager* as they prepared the prepackaged meals on the radiator of the engine. The lack of room for a waste disposal system meant they stored solid wastes in plastic bags in the wing section and vented urine through a tube in the fuselage. As Jeana said when *Voyager* landed on December 23, 1986, after nine days, 13 minutes and 44 seconds of flight, "If it was easy, it would have been done a long time ago."

It was inevitable that the press compared Jeana Yeager with Amelia Earhart, and Dick Rutan, with Charles Lindbergh, both were thin and near look-a-likes. They were modern day aviation pioneers and just as newsworthy. The flight illustrated the progress women have made. Unlike Earhart, Yeager was newsworthy not because she was a woman, but because her achievement was newsworthy.

"Dick and I more than doubled the world record when we flew non-stop and non-refueled around the world - the first time it had ever been done." (They had flown 26,358 miles, starting out with 1,200 gallons of fuel and landing with 18.3 gallons left.)

"It was exciting watching it all come together, exploring your own self and finding out, 'yeah, I can do this; I'm capable.' It was a fun discovery period." The *Voyager* was one of those rare opportunities that hardly ever comes along in a person's lifetime. "When you have that opportunity," Yeager said, "it's hard to say no. I certainly couldn't."

In May 1987, Jeana Yeager was awarded the Collier Trophy. Although awarded since 1911, Jeana was the first woman to win the most prestigious award in aviation. In April 1992, Jeana Yeager also received the coveted Crystal Eagle Award, presented annually by the Aero Association of Northern California. Previous recipients included General James Doolittle, Chuck Yeager, and Bill Lear.

Yeager's philosophy is focused. "All the experiences of your life are training and developing for your next level of expertise. I would say that *Voyager* was a training simulation for my next level. I have no idea what the next level is going to be, but I know I'll be capable of doing whatever I decide to do." See Fig. 7-9.

Jeana Yeager would like all young women thinking of a career in aviation to: "Believe in your dreams. Anything is possible."[22]

Chapter Eight

Federal Aviation

If you can't see any opportunities where you are now, don't waste your time criticizing the darkness... Light a candle to find your way out. Arlene Feldman

New laws and regulations that mandated affirmative action programs brought an influx of women employed by the federal government into aviation professions. In 1971, Ruth Dennis became the chief of the FAA's San Diego Flight Service Station, and on May 16, Gene Sims became the tower chief of the Cuyahaga County Airport in Ohio. Both were the first women ever to hold these positions. At the end of the decade, the FAA rewrote its Aviation Career Services pamphlets to include only gender-neutral terms; for example, ramp serviceman became serviceperson, and "he" and "she" were used jointly throughout the text. In addition, the FAA also reprinted for public distribution a series of magazine articles about women. The four reprints included information on women in Navy and Air Force aviation, as well as more general surveys. Yet, for all the official "equality," numbers and attitudes changed slowly. During the late 1970s women accounted for only about five percent of the total number of air traffic controllers. Lynne DeGillio, who was an air traffic controller at John F. Kennedy International Airport in New York, described the tower as retaining its "locker room ambience" despite the addition of women.[1]

Today, more than ever, more women serve in a multitude of capacities in the federal government. Elizabeth Dole, appointed U.S. Secretary of Transportation in 1983, was considered the most influential individual in American aviation. She had overall responsibility for federal aviation programs as well as those in the Coast Guard.

Arlene Feldman - FAA Regional Administrator - New England Region

Arlene Feldman has been an ambassador for aviation for more than a decade. Starting with her sponsorship of the 1983 Airport Safety Act, when she was Director of Aeronautics for New Jersey's Department of Transportation, to her current position as the highest-ranking non-politically-appointed woman in FAA history, Arlene Feldman has been sharing, with everyone who will listen, the mission of the FAA and the role of women in aviation.

Feldman combined her favorite hobby and education to literally move upward in the FAA and into the ranks of women speaking out for aviation and women. After graduating from the University of Colorado with a degree in political science (cum laude), she went on to earn a law degree at Temple University. At law school her advisors suggested she combine her interests in flying with her studies. Her first job with the United States Railway Association kept her on the ground, but not for long. In 1982, she got back into the air when she became New Jersey's Director of Aeronautics.

Some of her early accomplishments were to direct and implement a multi-grade educational program to promote interest in aviation, and establish a state and regional disaster airlift program.

Feldman's success at her first aviation job led to more challenging jobs. Her next job was as Acting Director and Deputy Director of the FAA Technical center at Atlantic City International Airport. There she managed more than 1,200 scientists and a 5,000 acre research and development test base. Some of this research led to improved safety regulations for seat and interior panel materials, evacuation slides and emergency exit lighting. See Fig. 8-1

Success breeds success and for Feldman this meant another challenging assignment as Deputy Director of the FAA Western-Pacific region in Los Angeles. There she managed more than 5,300 employees in the FAA's largest region. She served as principal spokesperson to the Los Angeles media, and she handled some difficult press conferences.

Today Arlene Feldman is the New England Regional Administrator for the FAA, and admits that there was a time when the FAA would never have dreamed of having a female regional administrator, much less a female deputy administrator. These jobs didn't happen by accident or coincidence either. Her experience was crucial and her education was just one of the tools she used to advance in the once male-world of the FAA.

Because of her education, Feldman could challenge the people who pointed to her and said she hadn't paid her dues. But she doesn't have to answer their challenge. Her record speaks for itself. She attended law school full-time, worked full-time, raised a family at the same time and THEN passed the bar. "It is very easy to criticize what others "don't have," she says. "Unfortunately most people judge others by their actions and themselves by their intentions." Her point is her training and education were attained through hard work and perseverance, she didn't think about what the other person was doing. She saw what she needed and went for it.

Fig. 8-1 Arlene Feldman

The future for women in the FAA wasn't always as bright as it is today. In post World War II America, many women in aviation could not find FAA jobs, and when they did, found themselves unable to move from the Flight Service Stations where they were placed. One of Feldman's early mentors was one of the finest weather briefers she's ever known. "When it came to a weather briefing, no pilot cared whether or not she was a female. They went to her time after time because she was good. However, there were limited opportunities in those early days of the FAA for women. Even so, this woman did not let that affect her performance."

The list of high-ranking women in the FAA is growing everyday. If women are to continue to make progress in the FAA, Feldman feels they need to develop certain skills, and they are not "female" skills. They are based on demand. She makes the point that women need the same skills as males do. Computers are the first and logical step. From the lowest levels and upward, the FAA is a computer driven organization.

There are other skills that are equally important. A pilot's license for example. "We're in an environment where our ability to perform is a function of our knowledge of the system. We need to know - we *must* know - what's going on in the system. If your desire is to excel, to reach a plateau, then you need to become a pilot. I am a pilot and because of that, I have a perspective of the system that a non-pilot simply cannot have."

Perhaps the most important skills, and the ones which Feldman thinks are the most ignored, are the interpersonal skills of oral and written communication. These are Feldman's most important tools at her disposal. She says, "I've read that one of the marks of real intellect is the ability to communicate with the genius and the fool. If you can't *tell* somebody what you know, it doesn't *matter* what you know."

Arlene Feldman has some sound advice for young women thinking of a career in the FAA. "You've got to make a name for yourself. Get your degrees. Get your ratings. Be sure you are in the spot where you need to be." Recently someone showed Feldman a quote from a sports magazine about an up-and-coming golfer. "All of us here have the same dreams," it said. "The only difference is some work harder than others." That golfer was just seventeen years old.

One would think that Feldman has proven herself and there should be no question of her competence. "People still challenge me about what it is I am and what I know. They say, 'she just got there because she wears a skirt.' My response is no response. Whether or not a skirt got me here is immaterial. The main point is that wearing a skirt is not what keeps me here.

"The point is," says Feldman, "that making the grade is not the point. The point is *maintaining* the grade. You've got to push and push hard. Nobody's going to say, let's promote a woman today. You will see very few help wanted listings for "WOMAN,'" Feldman says. "If you can't see any opportunities where you are now, don't waste your time criticizing the darkness. Light a candle to find your way out.

"More than 400 years ago in *The Prince*, Machiavelli wrote, *It must be remembered that there is nothing more difficult to plan, more uncertain of success, nor more dangerous to manage than the creation of a new order of things.* "He knew then what we know now: The good old boy network isn't going to dismantle itself just because we don't like the status quo. What's the answer? The answer is production, and quality and perseverance. And preparation."

Laura Goldsberry - Federal Officer

In 1986, the U.S. Customs welcomed its first female pilot. Perhaps the word "welcomed" is too strong, because in the years that followed the attitudes have not mellowed much. Laura Goldsberry was 18 years old when she entered Purdue University's Aviation program. She had only been up in an airplane once but it was like a comfortable shoe. She knew that flying was what she wanted to do for a career. While it was easy to say "I want to fly," it was far from easy for Laura to turn her dream into a reality - but she did.

Laura put herself through college and worked three jobs to afford her flight instruction and her ratings. After graduating from college, Goldsberry held the usual flight instructor and charter jobs, both building her time and honing her skills. Then came the first break for Laura. Purdue University hired her as their first female flight instructor in the General Aviation Department. For the next seven years Goldsberry taught and during that time, earned her helicopter rating. It was then that she suggested to the university that they begin a helicopter training course. But like many of the women in this book, she was a bit ahead of the traditional thinking. The university politely declined her suggestion.

In 1986, Laura Goldsberry came face-to-face with the ugly side of her chosen career, when the U.S. Customs hired her as their first female pilot to fly the Cessna Citation jet and UH-60 Blackhawk helicopter. Some of her fellow pilots for some inextricable reason treated her as if someone had handed her flight training and ratings. This was sad and truly ironic, since many of her fellow pilots had received their flight training from Uncle Sam, as military pilots.

Aside from that one aspect of her career, Laura says of her job, "Life as a federal officer/pilot definitely has its interesting moments. Unfortunately, most of the exciting details are still classified information right now."

Laura's job takes her around the country as well as overseas and she often works undercover. She said recently on reflecting back over her career, "While every ounce of my career has been a struggle, I wouldn't trade it for the world. I live to fly - catching crooks is a bonus."

Air Traffic Control

Right after World War II began, women were called upon to handle many different jobs that only months before were strictly "men's work." One of those jobs was that of an air traffic controller. In 1942, fifty women began training as Army radio operators at New York's LaGuardia Airport. These women were the first women used as air traffic controllers. During the war years, women filled hundreds of air traffic control jobs in the Army and Navy but after the war, things went back to the old ways. The women were told that they were only needed for the emergency war situation. They were now to go back home and raise their families and make room for the men.

After the war, aviation boomed, and the industry could not keep up with the growth. Ironically, women with aviation and air traffic control experience found few jobs with the government. Those who worked as controllers during the war would have to reestablish their credibility to be accepted by their male co-workers. The common perception was: "Women could not handle the stress and pressure of the job. Women were too emotional, and delicate to be air traffic controllers. It was a man's job."

Inroads were made and by 1983, about eight percent of all air traffic controllers were women. Many of these women belonged to an organization known as the Professional Women Controllers (PWC), which was founded in 1979. PWC originated as a dream of Jacqueline Smith and Sue Mostert when they met at the FAA Academy in 1968. There were many problems faced by women in this profession, principally because of their isolation. It was, and still remains, common for a female controller to be the first and only woman at a given location. Smith and Mostert wanted to create a women's organization for air traffic controllers as a network of communication that could break down the barriers of isolation. Thus, it was not created to compete with the union functions of the Professional Air Traffic Control Association. Naturally, PWC's membership was affected by the 1981 strike, which led to the disbanding of PATCO, but PWC was not forced to break up. The new union, the Air Traffic Control Association, is now an important supporter of PWC's activities, which include an active recruiting program to encourage women to enter the profession. The goal is to help women like Dayle Buschkotter realize their full potential in the profession through education, communication, and support.

Dayle Buschkotter - Air Traffic Controller

Dayle Buschkotter's first experience in a small plane was on a blind date. She got to sit in the right seat and the experience changed her life. She began earning her private rating while in college. Her mother wasn't crazy about the idea. "Flying is for rich people," she said, "and it has no career prospects for women." This was around 1977 and women were beginning to gain access into the aviation industry, and raising their own profiles.

"My mother also was the first to recognize my achievements," said Dayle. "On the day of my check ride, in 1977, she was my first passenger."

From that point, Dayle wasted no time building her ratings. Within two years she had her commercial and instrument ratings and had a job with the FAA. By this time also, the government had begun to nudge the business sector and the FAA was actively "looking for a few good women" to enhance their ranks.See Fig. 8-2.

Dayle's first air traffic control job was at Palmdale, Calif. and since then she has worked at towers at some of the west coast's busiest airports. In 1981, she became a supervisor at Orange County Airport.

Following that promotion she earned her CFI - airplane and in 1982, earned her rotor rating and became a Whirly-Girl. She has flown various fixed and rotorcraft including the Bell, Enstron, and her favorite, the T-34.

Fig. 8-2 Dayle Buschkotter

Dayle feels one of her major missions in life is bridging the gap in knowledge and communications between pilots and air traffic controllers. "I am confident that I can make a positive difference to this extremely important part of the aviation industry," she said. "Sometimes I don't see the immediate impact of my efforts but I know that if I continue to do positive things, good things will come of my efforts."

Some of Dayle's efforts include facilitating pilot/controller workshops, controller workshops, and she talks to the airlines through the NATCA. She deals with general aviation through FBOs and special invitations. Dayle is also an accident prevention counselor and has been on the planning commission for Helicopter Awareness Day since 1985.

Dayle is also the international vice president for the Whirly-Girls, past vice chair/membership for the Long Beach Chapter of the Ninety-Nines and past president, secretary-treasurer of the Professional Helicopter Pilots Association of California. She is also on the editorial advisory committee for *Rotor* magazine. In her spare time, Dayle is a mega warbird fan. Each year she helps sponsor a T-6 at the Reno Air Races.

Andy Edmondson - Air Traffic Controller

Andy Edmondson started flying lessons in 1970 in Morristown, TN and remembers being in awe of the controllers she heard when she flew in and out of Knoxville. "I thought they were like gods," she recalls. "My flight instructor, Evelyn Johnson, even suggested that I look into air traffic control but I thought she was crazy."

By 1974 Andy was still building her time, but trying to eke out a living flying bodies and canceled checks was tough, so she took the air traffic controllers test and was accepted into the agency in 1975. "I was the only female in a class with eight men. Everyone, except for me, had been a military controller but I was the only one with a pilots license. I think it was my aviation background that got my foot in the door."

Andy first checked out as a controller at Memphis center and later transferred on to the Boston and Greensboro facilities. Today she is controlling traffic in Charlotte, NC, the 7th busiest airport in our nation which handles more than one hundred operations an hour. Most outsiders assume a controllers job is stressful but Andy believes that the stress level is relative. "When I first dropped my daughter off at the baby-sitters I couldn't get over the mass confusion with little children running all over the place and thought to myself, now that's a stressful job! You have to like doing a job like this, I think it's more of a challenge than anything else."

Common in most occupations within the aviation industry is the pressure women feel to be better than their male counterparts in order to survive. Scrutiny causes this tendency and seems to be inherent regardless of the profession and air traffic controlling is no exception. "People will always be looking over your shoulder to see how you're doing and for me it's a matter of pride. You want to be good, but you don't want to make a big deal out of it. I know in my heart I'm a good controller but I'd never say that to anyone. Any guy looking over my shoulder will have to see that for himself. There is a professional organization for women controllers but in all these years I have never joined because I don't feel we should separate ourselves as "women controllers" and "men controllers. We all do the same job regardless of the gender."

Andy's husband is also a controller and they have had to make many sacrifices in their careers to stay together. Their first transfer from Boston sent Andy to the Greensboro tower and her husband to the Raleigh tower so they bought a house that was halfway between both airports and commuted in opposite directions each day. Presently both have transferred to the tower in Charlotte and although they are eligible for supervisory positions, a longstanding FAA policy prevents spouses from supervising spouses. Opting not to be promoted, they are content to remain at the controller level in order to stay together in Charlotte.

Andy's successful career in air traffic control has endured eighteen years and she admits that with retirement looming she is starting to see the light at the end of the tunnel. For now, her "escape" from the rigors of air traffic is their 26 acre country hide-away outside of Charlotte. "When we were looking for a home out here, we just told the Realtor that we wanted something as far away from the airport as possible!"

Wally Funk - Former NTSB Investigator

On December 9, 1974, Mary Wallace Funk II, "Wally" was the first female in history to become an Air Safety Investigator for the National Transportation Safety Board (NTSB). The NTSB is the smallest government agency in the United States and answers only to the President and Congress in its quest to investigate aircraft accidents. It was touted as an unprecedented accomplishment for women, but it was a mere entree into the impressive list of aviation "firsts" and achievements that Wally had already attained. See Fig. 8-3.

There was no doubt that Wally was capable of succeeding in a male-dominated agency as she had already become the first female Federal Aviation Administration (FAA) Inspector three years earlier. She was accustomed to national attention and was no stranger to the acclaim and limelight these jobs attracted. As the "women's-lib" era of the seventies provoked radical change in the work force, Wally held on to the belief that a woman should have a place in any industry or field that interested them.

Wally grew up in a non-aviation family in the snow skiing capital of Taos, New Mexico and was more interested in model airplanes than dolls even at an early age. As a teenager, Wally was a distinguished rifleman and nationally ranked downhill and slalom skier. It seemed evident that Wally would master and excel in everything and anything that held her interest.

By 16, Wally had entered Stephens College in Columbia, Missouri to start her flight training. She graduated in 1958 with an Associate of Arts degree and was rated first in her class of 24 pilots. In 1964, the college recognized Wally for her aviation accomplishments and made her the youngest woman in the history of the college to receive their distinguished Alumni Achievement Award. Wally completed her ratings and received her Bachelor of Science Degree from Oklahoma State University, capturing top honors as, the "Outstanding Female Pilot," and the "Flying Aggie Top Pilot." She won the "Alfred Alder Memorial Trophies" two years in a row.

Wally propelled herself into the aviation industry at the age of 20 by landing a job as the first woman flight instructor at Fort Sill, Oklahoma. As a civilian flight instructor, she taught flying to non-commissioned and commissioned officers of the United States Army. During her employment there she soloed more than 400 service men and sent a total of 500 on to private, commercial, instrument and air transport ratings.

In February 1961, Wally was one of twenty-five women candidates secretly invited to undergo preliminary astronaut testing. In what seemed to be the United States' attempt at being the first nation to put a woman into orbit, Wally and the women endured the same grueling and demanding tests that the

Fig. 8-3 Wally Funk

Mercury astronauts had completed. Tests included swallowing three feet of rubber hose, drinking a pint of radioactive water, hours of seclusion in a sensory depravation tank and altitude chamber test experiments to name a few. Wally scored amazing results in each phase and surpassed John Glenn in two of the physical tests. In the sensory deprivation chamber, Wally set a record, remaining in the chamber for 10 hours, 35 minutes. She could have stayed on, but the officials stopped the test.

In February, 1962 Wally completed Phase III of the tests and was the first woman at El Toro Marine Corps Base to undergo the Martin-Baker seat ejection test and the 39,000-foot-high altitude chamber test.

The test results concluded that women were as capable and as suitable for space as the men but in 1962 the program was suddenly halted with no conclusive explanation. "Naturally I was crushed," recalls Wally, "I'd have given my life for the space program. We served as human guinea pigs in a sense, but in doing so I believe we served a definite purpose and substantially aided the male-dominated space efforts."

In 1970, Wally submitted her fourth and last formal application to NASA. Again they rejected her application but advised her that if she could get her engineering degree in 12 months they would reconsider her. Needless to say, that was impossible.

Years later Sally Ride said that the first women astronaut candidates made her flight possible by setting physical and psychological base lines for her and other women to launch into space.

In the years that followed, Wally embarked on an extensive three year tour as a Good Will Flying Ambassador. She covered 50 countries and 80,000 miles throughout Europe, Africa, the Middle East and Russia. She longed to meet with Russian cosmonaut Valentina Tereshkova and tried to arrange a meeting with her while on tour. Disappointingly, the meeting was not allowed by the Russian government citing that the two nation's competitive race-for-space prevented such a friendly exchange.

Back at home, Wally's inexhaustible energy drove her to explore nearly every area in the aviation industry, leaving few props unturned. She flew as a flight instructor, a charter pilot and a chief pilot and taught ground school and aeronautical science classes to thousands of students. She competed in many air races including the famous Powder Puff Derby and the Pacific Air Race where she captured the first place trophy in 1975. She was the 58th woman in the U.S. to earn an air transport pilot rating and has done everything from parachute jumping and hot air ballooning to soaring and hang gliding. She has been a member of the Ninety-Nines for 35 years, and is a Colonel in the Confederate Air Force (CAF). Wally has 13,400 hours of flying time.

Opportunities for women in the military had not yet become a viable route and although women had not yet been hired by the airlines, Wally submitted her application with hopes of employment. "I was finally granted an interview with one major carrier. The personnel director examined my qualifications and told me that I was head and shoulders above his male applicants, but because there wasn't a women's restroom in the flight department they couldn't consider hiring me."

Wally decided to put her talent and skills to better use by serving in the public sector. In 1971, she became the first woman to successfully complete the FAA General Aviation Operations Inspector Academy course which enabled her to survey accidents and certify pilots. A promotion in 1973 earned Wally the distinction of being the first female in the United States to become a specialist in the FAA System Worthiness Analysis Program. This task force audited and evaluated the level of instruction in all FAA-approved flight schools and air taxi operations. This was Wally's avenue to change the long-standing FAA attitude of "violate then educate" to the more appropriate adage of "educate then violate." See Fig. 8-4.

The NTSB wooed Wally away from the FAA in 1974 and she became the first female Air Safety Investigator. "The training that I received for the position was the best aviation education available. Back then investigators were sent to study with the companies of every aircraft, engine, propeller, and component manufacturer in the United States and abroad to learn accurate systems design

Fig. 8-4 Wally Funk

and operations. Once on the accident site, we were able to take all this aircraft knowledge and information with us to begin the accident analysis process.

"The first accident I was called out to investigate was a mid-air in the Los Angeles area which occurred over a school ground. The entire office turned out to see how I'd handle the case and frankly it was a very traumatic experience for me. While all the training I had received focused on the mechanical process, nothing had prepared me for the psychological impact of an accident scene."

In a graphic sense, an accident scene can be bloody, chaotic and emotionally draining. Having to handle shocked survivors, lifeless corpses and hounding reporters, was more than Wally was prepared to deal with. "Images from that first case stuck with me for a long time and it took about 6 months before I could easily adapt." Wally outlined the need for psychological training in the educational process and appealed to her superiors for implementation. Wally herself contributed to the program's development and today investigators have her to thank for being better prepared for the catastrophic impact of field work. "As an investigator at the scene of an accident I performed all phases of the investigation including custody over aircraft wreckage, mail, cargo and bodies. I established wreckage site security, questioned witnesses and photographed appropriate aspects of the wreckage. I arranged for wreckage recovery, transport and storage and ordered whatever tests would be vital to the investigation including, engine tear-downs, failed parts analysis, and autopsies. I also had the authority to arrest anyone on the premises who interfered with my investigation. Often I reserved this threat for the newspaper reporters who hounded me to solve the accident in time for their 6:00 newscasts."

Wally and her judgement were well respected in the agency. She used methodical and meticulous procedures to solve her cases and was not afraid of getting her hands dirty. With manufacture's specifications in hand, Wally would spend hours painstakingly dissecting aircraft pieces like a jigsaw puzzle, to search for accident evidence. She also conducted exhausting research through aircraft maintenance and pilot records to look for human elements that might have contributed to the cause. "You couldn't be satisfied to establish just the primary cause of an accident because few accidents result from a single factor. Usually it's a sequence of events. I found the single best attribute that an accident investigator can possess is an open mind. You have to approach accidents without preconceived ideas so that pertinent evidence isn't overlooked."

Wally's creative techniques in investigation even included enlisting the help of psychics on two occasions to help pinpoint the location of missing aircraft, both with positive results. Her efficient and thorough investigative methods helped solve 350 accidents she handled in her 10 1/2 year tenure. "Solving these accidents helped us learn how to prevent another. It was rewarding to put forth new recommendations and regulations to keep accidents from repeating themselves."

Deregulation and the Carter administration brought changes to the agency. Investigators found themselves stuck behind desks with mountains of paperwork instead of out in the field with their sleeves rolled up, solving cases. "There was so much I had learned from accident investigation. Flying is still the safest form of transportation. Eighty-seven percent of general aviation accidents are caused by "poor flight training" so I thought there was no better time than the present to go out and use my experience to educate the general aviation public."

Wally retired from the NTSB and promptly designed a seminar called "How to Fly and Stay Alive."

Today Wally maintains a steady course on the lecture circuit, educating pilots on how to put safety and common sense into flying. Wally stresses that safe flying is no accident and that airplanes don't cause accidents, people do. From a former investigator's point of view, "There are no new accidents, just new pilots trying the old ones."

Wally stresses the importance of proper pilot attitude and preflight planning, and how to determine a go-no-go situation and the fatal results that the two main accident contributors; "got-to-go itis" and "loss of control" have on flights. Attention is also focused on inflight weather hazards, emergency procedures and survival equipment. Her captivating presentations include more than just a recitation of facts and figures, she also introduces creative techniques to enhance safety and heighten awareness. Among these are gems like the "Wally Stick," a simple wooden dowel that can detect the slightest internal imperfection in a common propeller blade.

In addition to the hectic schedule of speaking engagements, Wally practices what she preaches as a flight instructor and devotes her time and energy as a national judge for NIFA (National Intercollegiate Flying Association). She also assists the FAA in safety counseling as well as in the development of cockpit resource management courses.

Success at so many levels has made Wally a highly visible force in our industry. Her carefree spirit captures the heart of all who meet her and her contagious love for flying seems to spark an aviation interest in all that listen to her speak. The flying public is indebted to Wally Funk for the lifetime of service she has provided in her quest for an accident free environment.

Pam Kleckner Sullivan - NTSB Investigator

Because each new Presidential administration appoints a new Chairman to the NTSB, reorganization and restructuring breed change within the agency. These changes can cause a shuffling of staff and personnel, but one thing remains the same, and that is the vast number of opportunities available to women in the NTSB today. Co-op programs like the one established with Embry Riddle University introduces the vocation at an early stage and allows a hands on approach to those interested in transportation safety.

In 1977 Pam Kleckner Sullivan was one of just two women enrolled in Kent State University's aviation program and was pictured in her college year book along with those in "NonTraditional Jobs" because of the male-dominated aviation field she had chosen. Completing each new flight rating, Pam dreamed of becoming an airline pilot, that is until the Chief of Chicago's NTSB office came to lecture to the students about accident investigation in 1982. "After he spoke, I was asked to fly him onto Hopkins so he would make his connection. We talked and he encouraged me to apply for an investigator-trainee position that was opening up. I figured it had to pay more than flight instructing so I pursued it"

Pam was hired by the NTSB in June of 1983 and has sustained a fulfilling career in air safety. She has spent ten years between the Washington and Chicago offices and has been promoted as the only female unit supervisor in the agency. Though the position is basically rooted at the management level, Pam is still able to balance the role with her share of field work.

"Challenge is the only reason to stay in a job like this" remarks Pam, "It's very demanding to be on call 24 hours a day, weekends and holidays, and crash sites have us trudging through baking hot corn fields in the summer and blowing snow and cold in the winter. Emotionally the accidents still bother me but when you're out in the field you have a job to do and you're too busy organizing the investigation and attending to details to let it effect you. It's when you get away from that atmosphere and have to deal with the human aspects of the accident like, informing the next of kin or reviewing flight data recordings, that it hits you."

As for pointing fingers and laying blame where accidents are concerned, it is far from a priority as real job satisfaction comes from approved changes the investigators propose to improve flying safety for the public it serves. "After we perform the field investigation and complete a factual report of the accident,we send it on to headquarters and they make the final determination of probable cause. As the NTSB, we can't mandate changes in the industry but we can spearhead campaigns for proposed changes that we feel are most important."

Pam is still in-tuned to the airline industry she was once attracted to because her brother is a pilot for a major carrier, but comparing "job security" she is glad to have chosen a career with the NTSB. She is comfortable as a part of our industry's puzzle that provides the solutions and hopes to continue moving up the ladder, possibly to regional director and beyond.

Catherine Roetzler - Leadplane pilot - US Forest Service

For Catherine Roetzler, airplanes seemed to be a way of life from her first recollections. She was raised on a farm about four miles from an airport and airplanes always seemed to be there. They were part of the sky. But in her growing years, she never thought they would become such a big part of her life.

For Kt, as she is known, her senior year in high school brought her first major decision in life. If she wanted to live at home after graduation, she would have to go out and get a job. The only place that came to mind was the airport. She went down to see the airport manager and found out that the restaurant needed a waitress, the National Weather Service needed a secretary and the car rental agency needed someone, too. She took the civil service exam and went to work for the weather service two weeks after graduation.

One day her life took a sudden and unexpected turn. A flight instructor came in for a weather briefing. In the chit-chat that followed, he asked her if she had ever tried flying. Kt responded that she was just a high school graduate and didn't think she was qualified. The instructor took some of her misconceptions away that afternoon. He told her flying was for anyone who wanted to fly. She could do it if she really wanted to. Kt thought about it and went home and discussed it with her mother. The next day, June 21, 1978, Kt will never forget. "I took my first lesson and I knew I was going to be a pilot the rest of my life." See Fig. 8-5.

Kt enthusiastically began her flight lessons, but certain parts of flying didn't come easy to her.

"The hardest part of flying was learning to land." she said."I trained in a Beechcraft "Musketeer" and I would always thank God for the sturdy landing gear."

Kt earned her private license on August 20, 1979, but not before she experienced many frustrations and discouragement.

"I remember the first time my instructor got out of the airplane and said, 'take it around the patch three times.' I was so scared. I remember on the downwind not wanting to land - but I knew I had to. This landing was really bad. I bounced the Musketeer to a stop. I taxied over to my instructor knowing he was going to be upset with me for having done such a poor job. He made me go

Fig. 8-5 Kt Roetzler

around the pattern two more times. I bounced the plane into the ground both times. I was very discouraged and it was then that I decided to give up flying. He wouldn't hear of it and talked me into staying with it."

Working the weather service was convenient for Kt. At lunch time, instead of having lunch, she would shoot touch and goes.

"It was still discouraging but I would not give up. Then one day I made such a smooth landing that I did not even know I was on the ground. A light in me clicked on. I had made a nice landing! I made a lot more bouncing landings but they became fewer and fewer."

After Kt got her private license she was offered a job as a secretary in a little town with an FBO. She again used the convenience of being situated at the airport to advance her skills. She quickly proceeded through the instrument, multi-engine and commercial ratings. At that point she realized there was no future in her secretarial job and besides, she had developed a new skill that beckoned her. It was time for Catherine Roetzler to move on and up, into the sky.

"I moved to El Paso and freelanced. I flew anything, anywhere, anytime. Because I didn't make enough money flying I took a night job as a cocktail waitress that paid more than flying."

During her freelancing, Kt did get a chance to fly a number of different singles and light twin aircraft. She also began to meet people who would later help her career.

At a time when many of her friends were getting jobs with the airlines, Kt thought about it and decided that airline flying did not appeal to her.

In October, 1984, Kt made another move. This time it was to Tucson, Arizona, flying freight. The following three years held new learning experiences, too. She flew freight all over Arizona, Texas, New Mexico and parts of California.

"I flew for different companies and learned more about airports, weather and back-to-basics flying in those three years than I had learned up to that point.

"One company I worked for was pretty shady. There was never a time when I could remember everything on the airplane working at the same time. There was always some kind of fuel line, mechanical, electrical, hydraulic, gear, flaps, etc. problem. Sometimes a combination of problems. I looked at this as a positive experience. I've had eight engine failures in 14 years of flying. I'm used to system failures and now when something goes haywire on an airplane I know how to deal with it."

Today Kt still calls herself "an ol' freight dog."

"Flying freight was hard and dirty work. But it was also the most fun I've ever had flying - except for my present job."

In June 1987, Kt went to work for the U.S. Customs. It took only two weeks for her to realize she had made a wrong choice. But she hung in there until April 1989 when she landed a job with the U.S. Forest Service. "It was here that I found my niche in life," she said.

"The Forest Service is a fine agency to work for. It's a place where most of the people spend their time with Mother Nature."

Kt's job as leadplane pilot is to help coordinate forest fire control. Her role is important, as she is responsible for making the process of putting out wild land fires from the air as safe as possible. This takes the cooperation and teamwork of everyone involved, from the helicopter pilots to the tankers and dispatchers to the people on the ground. "As you can imagine," she says, "this career is rewarding, satisfying and quite a challenge.

"I've been flying for 15 years and have never regretted any of it. I never really sat down and thought out what I was going to do with my flying career but I do know that my flying has never let me down, and never said no to me. It is a thrill every time I start the engines.

"I never thought of myself as a female pilot, just a pilot. I never said no to a job unless it was illegal, or went against my morals and values. I found out early in life that being yourself, being honest, and working hard will pay off in more ways than just monetarily."

Chapter Nine

Aviation Education

Aviation is still considered a man's world by many. The time to reach young ladies is during their first years of school. Research has shown that although children may change their minds several times about their eventual careers, the possibilities of them selecting a non-traditional role must be nurtured at an early age. Dr. Peggy Baty

While Harriet Quimby was probably the first woman journalist to write about women in aviation, the practice fortunately did not die with her. Today that tradition is alive and well, and yielding positive results. One of Amelia Earhart's legacies is her reputation for promoting women in aviation. At various times in her life, Earhart was the aviation editor of *McCalls* and *Cosmopolitan* magazines. She was often quoted in newspapers and lectured frequently about fostering a positive attitude about women in aviation.

Dr. Peggy Baty - Educator

"You are limited only by your dreams and your willingness to see them through." That's Dr. Peggy Baty's view of life.

Dr. Baty decide to become a pilot before it was fashionable for women to pursue this career. It was in college that the now infamous flying bug bit Baty. She was in the right seat of a low wing 1946 Ercoupe when it happened. "The pilot let me sit at the controls and it was like nothing I had ever done before. It made me reconsider all my career plans. I was studying to be an elementary school teacher but wasn't sure I wanted to teach kids. Flying made me see something else." see Fig. 9-1.

Baty wrote to her mother of her intent to join the aviation world. Her mother was enthusiastic but misunderstood her daughter. Peggy received some clip-

Fig. 9-1 Dr. Peggy Baty

pings from her mother that described the benefits of being a flight attendant. Peggy responded back that she had no intention of sitting in the back of the plane. She was going to be a pilot! Maybe even a captain someday. She took a year off from her school to earn her ratings.

Slowly the realization of the real world of flying crept into Baty's consciousness. "I flew into an airport in Georgia with a female student and the gentleman who ran the fixed base operation asked, 'Where's the pilot?'"

"When I would fly with a male student on a cross country flight, I could count on the lineman walking up to the student as we secured the plane and asking if he needed any fuel, etc.

"Then too, when people asked me what I did for a living, and I replied I was a flight instructor, the standard refrain would follow.'Oh, and do you fly too?' I guess they thought I taught flying in the classroom."

One of Baty's greatest challenges was to overcome this misconception. Sometimes she was successful, sometimes she wasn't. While teaching flying as an independent contractor in Jasper, Tennessee, Baty was introduced to one of the students as the airport's new flight instructor. "The man looked at me, took one step back and said,'I never flew with no woman before.'" To counteract this, Baty offered him an hour of free instruction and asked him to then decide on her qualifications. After that hour, he became one of her best boosters, encouraging others to come to the airport and fly with the "lady instructor."

On the other side of the bias coin, a young woman showed up at the Jasper Airport one day inquiring about flight instruction. When she was told her flight instructor was a female, she changed her mind, preferring a male instructor.

Baty came down from the skies long enough to earn a masters degree in aerospace education and her doctorate in educational administration and supervision at the University of Tennessee. Baty soon moved on to chair the administration department at Georgia State University in Atlanta. Her next move was to Embry-Riddle's Florida campus in 1986. There she was promoted to Associate Vice Chancellor/Associate Dean of Academics and relocated to their Prescott campus. In 1991, Dr. Baty became Associate Vice President and Dean of Parks College of St. Louis University. With this appointment Dr. Baty became the first woman to head Parks College (the oldest aviation college in the country) and the first woman to head an aviation college anywhere.

"Typically women don't consider aviation as an option. Not because they don't think they can do it. It's just not something they are exposed to. I think once women are exposed to aviation and have an opportunity to see women role models they will realize they can be successful in this field. Education is the key."

Dr. Baty doesn't just talk about it, she takes action, and is a strong proponent of women in aviation. She is the author of numerous articles on aviation education and women in aviation, and has presented papers on women in aviation to various colleges, and the General Aviation Manufacturers Association.

She was the co-director of the national "Images of Women in Aviation - Fact vs. Fiction Conference in 1990, Director of the second annual national Women in Aviation Conference in 1991, and the third annual Conference of Women in Aviation in 1992. Since the demand and attendance for these conferences has grown each year, Dr. Baty intends to continue them in the future.

Dr. Baty says, "Even today, many people do not realize that women are successful in the aviation arena as pilots, mechanics, air traffic controllers, airport managers, etc. When the general public is asked to consider a woman pilot, usually the only name that comes to mind is Amelia Earhart. Curriculum materials used in aviation education often show only examples of men who have been successful in this field. If a female example is given, more often than not, Amelia Earhart is the only one cited.

"Aviation is still considered a man's world by many," said Baty. "The time to reach young ladies is during their first years of school. Research has shown that although children may change their minds several times about their eventual careers, the possibilities of them selecting a non-traditional role must be nurtured at an early age.

"It is imperative that the aviation community play a role in the nurturing of young women today to be the aviation professionals of tomorrow," said Baty. "Clearly, the opportunities are there for expanded participation by women, but only if they receive the proper encouragement and exposure to this dynamic industry."

Professor Joan Mace - Educator

As Chair of the Department of Aviation at Ohio University, Joan Mace has brought her vast aviation experience and knowledge to a plane of higher learning. She succeeded Francis B. Fuller as Chairman after his retirement in 1985 and has been instrumental in the growth and development of programs that have put Ohio University on the short list of today's top aviation colleges.

Joan's aviation interests began during World War II as an inspector on the Hell Divers, (two-man Navy carrier planes) at the Curtiss Wright plant in Columbus, Ohio. Although she originally earned her private pilots license to be accepted into the WASP program, the war ended before she could be placed on active duty. In an era when opportunities were limited for women in aviation, Joan was determined to make it her career. She completed her commercial and instructor certificates and joined a flight training program that had been launched in 1946 at Ohio University to flight instruct returning veterans under the G.I. Bill.

As the only woman among 22 instructors and hundreds of students, Joan's nurturing nature earned her the nickname "Queen Bee." "I walked out to my J-3 one day and noticed something different about my airplane. As I got closer I saw that the guys had painted a big bumble bee on the side of my fuselage with the words "Queen Bee" written below." The name alone conjures up fond memories of the fever pitched days when 350 students buzzed the skies each semester from sunrise to sunset. Federal funding for the G.I. Bill was suspended in 1948 and the program ended. Joan married one of her fellow instructors and her career took a more domestic route. See Fig. 9-2.

An aviation department was established at Ohio University in 1963 and Joan was offered a position to return as an instructor. With her youngest of three boys in kindergarten, she accepted and returned to her first love of flying. Joan advanced her ratings and became one of just 60 women in the country to hold a multi-engine ATP, and was later appointed as a designated examiner for the FAA.

In 1968, the university requested that each of their instructors hold a degree, so Joan headed back to the classroom. She simultaneously taught, flew, managed a home and family and attended classes when she could fit them into her schedule. Joan finally received her own bachelor's degree, summa cum laude in 1978, graduating with her oldest son, who was earning his Ohio University degree that same year.

Joan has dedicated years of talent and service to the general aviation sector. She is the chief check pilot for the Ohio State Wing of the Civil Air Patrol where she holds command pilot status. She serves as a judge at National Intercollegiate Flying events and sits on the University Aviation Association scholarship and simulation committees. Each year on behalf of the All-Ohio Ninety-Nines, Joan presents the Arlene Davis award to the top female pilot at the NIFA National Safecon.

As a teacher and a scholar Joan is a productive member of the academic community and was promoted to the rank of full professor of aviation education. She developed an all new flight and ground school syllabus for the 141 school, and devised an alternative independent study program for commercial and instrument courses. Her presentations on "Ways of Increasing the Professional Status of Flight Instructors", "Innovative Teaching Facilities and Equipment" and "Competencies of Leadership" prove that her interests lie in improving awareness and increasing techniques to provide the best possible education for her aviation students.

Under Joan's leadership as the Chair of Ohio University's Aviation Department, a Bachelor of Airway Science Degree was established in 1986. This added curriculum increased enrollment and led to the completion of the new Francis B. Fuller Aviation Training Center which was dedicated May 11, 1990. To provide diversity for the students, Joan brought Alpha Eta Rho, the National Honorary Aviation Fraternity, to the campus. Professor Mace places great emphasis on improving job opportunities for her graduates and was instrumental in connecting Ohio University to the United Airlines Internship Program. "This is an innovative project that selects students each semester to take an apprentice role within the airline. I remember the days back in the 70's when airlines were hiring pilots right out of the flight schools. Today we have to better prepare the student for the competition they will face when they leave the training environment to actively seek employment within the industry."

Joan's administrative position has her flying more than just a desk. She continues to flight instruct, acting as the chief flight instructor for the university's FAA-approved 141 school and carries a load of at least three advanced students each quarter. She is the designated FAA pilot examiner for private, commercial, instrument, multi-engine and all the flight instructor ratings and gives stage checks and standardization rides to staff members.

Fig. 9-2 Professor Joan Mace

Each year the Alumni Medal Of Merit Award is presented to alumni who have achieved distinction in their chosen field. It is the highest honor that can be bestowed upon an alumni of Ohio University and in 1992 the award was presented to Joan Mace for her contributions to aviation and education. Of all Joan's accomplishments, her pride lies in the heart and soul she still holds with basic flight instruction. It is there the admired Queen Bee prefers to measure her success by the hundreds of fledgling pilots she has pushed from her nest.

Ann Lewis Cooper - Author

Ann Lewis Cooper, author of *Rising Above It,* the autobiography of pioneer aviatrix Edna Gardner Whyte, has spent almost a quarter of a century of her life involved in aviation. Ann decided to learn to fly when, as the wife of a fighter pilot, she was concerned to fly light planes with her family of five with only one aboard who knew anything about flying. She soloed and became a private pilot in 1969 and hasn't stopped learning. See Fig. 10-3.

A certificated Gold Seal flight instructor with an instrument rating (CFI&I) and an advanced ground school instructor (AGI&I), Ann has been an aviation instructor since 1971 and has been the chief instructor of three ground schools.

Fig. 9-3 Ann Lewis Cooper

She has been an FAA Designated Written Test Examiner and Accident Safety Specialist. At 2,000 flying hours, she continues to maintain currency, although aviation writing, historical research, and editing consume much of her time. She seeks to promote women whose contributions and achievements in aviation are and have been valuable. Believing that most school children, indeed most *people,* would answer with the esteemed "Amelia Earhart," if asked to name famous women pilots of the United States, she actively promotes female pilot pioneers and all women involved in aviation. Taking nothing from the achievements, contributions and selflessness of Amelia Earhart, Ann focuses on the less famous, but extremely daring and capable women who have helped to create the opportunities for the burgeoning role that women are finding in commercial aviation today.

Ann points out that although women were denied the opportunity to fly combat themselves, women of the United States taught military combat pilots for both world wars and, as members of the Women's Auxiliary Ferrying Squadron (WAFS) and the Women's Airforce Service Pilots (WASPs) women flew fighters and bombers during World War II. And although they were kept out of the cockpits of commercial airliners until recently, women were found qualified to teach male pilots to fly for the airlines.

While there have been inroads into these areas, Ann reminds us that today, women fly and instruct in airplanes, gliders, balloons, military craft and helicopters; they test aircraft; they build their own airplanes and airports; women compete in and win aerobatic contests; they pilot commuter airplanes and commercial jets; they participate in aviation law, aviation medicine and in government careers.

In her presentations to such organizations as the U.S. Army Military Academy at West Point, the Wings Club in New York City, and the EAA conventions, Ann introduces her audiences to a representative group, a few of the many pioneer aviatrixes that helped open the doors of aviation for today's modern achievers. She also speaks on "Writing for Aviation Magazines" and "The Art of Aviation Artists."

Ann believes that women who fly and who are interested in literature have great opportunities in aviation writing. In fiction and non-fiction, adult's and children's literature, short stories and screen plays, the list of possible aviation topics and personalities is endless. She says, "Although it isn't required that an aviation writer be an aviator, it certainly is advantageous. Flying has a language and a camaraderie that is unique. If flying itself is exhilarating, challenging and sometimes humbling, so to is writing about aviation and aviators. Tales of flying can be laced with danger and intrigue and they can include raucous joy and indescribable courage. Whether I interview pioneers or contemporaries, I find that an aviator's enthusiasm and pride is almost always contagious and that contributes much to the joy of this aviation writer."

Ann has published more than 600 non-fiction articles in national magazines, and is currently editor of the *Ninety-Nine News*, the magazine of the International Organization of Women Pilots. She also edits *Aero Brush*, the magazine of the American Society of Aviation Artists. Her ongoing series on the art of aviation artists is carried in *Private Pilot* magazine. Her first book, *Rising Above It*, was published by Crown Publishers, a division of Random House, Inc. in October, 1991. Her second book, *On The Wing*, the autobiography of aviatrix Jessie Woods, is under consideration and she is working on two screen plays, one on the WASPs and the other on the Jessie Woods story.

Married to Charles S. Cooper III, Major General USAFR (Ret.) the former commander of the New York National Guard and a division manager at Bellcore (Bell Communications Research), Ann lives with her husband in Berkeley Heights, N.J.. She has three children and four step-children. At this writing, they also enjoy five children-in-law and four grandchildren.

Carol L. Osborne - Author

Carol L. Osborne was born and educated in California, and for many years has had an intense interest in aviation. After graduating from Chico State University in California, she entered the field of aerospace. She worked for three years as a financial analyst in Ford Aerospace Corporation but she has been with Lockheed Missiles and Space Company for the past eighteen years. While most of her career has been as a logistics analyst of parts for the Trident Fleet Ballistic Missile Program, she recently moved into the Security Education and Awareness Media field.

Over the years Carol has developed an interest in our country's early aviation pioneers. Videotaping aviation pioneers for education and history is one of Carol's prime objectives. Aviation pioneer Bobbi Trout has been working with Carol on this project. They have worked on the project with only limited time and money for more than eight years. They know too that time is running out. Many of the Early Birds have already folded their wings.

The late Dr. David Hatfield (Aviation Historian of Northrop University) started videotaping several aviation pioneers thirteen years ago. Three years later he passed away, willing his video equipment and personal aviation memorabilia to Carol. It was through Hatfield that Carol met Bobbi Trout. Bobbi was retired, and interested in photography. See Fig. 9-4.

Osborne has devoted all of her spare time, including vacation time and weekends along with a great deal of her own money to videotape as many aviation pioneers as possible. "This is the best way I know of to honor Dave," she says.

Fig. 9-4 Carol Osborne

Carol estimates she has spent an average of $10,000 a year on the project. And this, she points out, doesn't include all the free help she gets from Trout. Expenses are high. Most of the people they need to reach can't come to them because of their age and are scattered across the United States. One Early Bird for example, Sir Thomas Sopwith, of Sopwith Camel fame, lives in England.

The two historians have driven thousands of miles to interview and tape many of the surviving pioneers. Today they have more than 250 in their library.

With all this information and documentation, Carol went looking for a publisher. Incredible as it may seem, she could not find any publisher interested. That's when she decided to start Osborne Publisher, Inc., in 1986, to publish two books: *Just Plane Crazy*, the biography of Bobbi Trout and her friends of the early 1920's and 1930's and *Amelia, My Courageous Sister.* Both have more than 500 photographs, documents and newspaper clippings, and were "a labor of love," according to Carol. It was more than just love, it was a passion. She believed in this project so deeply that she funded it at considerable expense from her own pocket. She even had to sell her car to make the last payment to the printer. But the sacrifice was worth it, as both books have gone on to rave reviews and sales are continuing over the years.

Carol had another aviation pioneer friend, Neta Snook Southern. Neta, who taught Amelia Earhart to fly, introduced her to Muriel Earhart Morrissey, Amelia's sister. The introduction grew into another friendship, and Muriel and Carol talked about updating and republishing Muriel's book about her sister. Carol had been collecting data about Amelia Earhart for more than ten years. The collaboration resulted in *Amelia, My Courageous Sister.*

"What I'd like to see in the future is a PBS program made from the tapes," said Carol. "But, now, the most important thing is getting to as many of the remaining Early Birds as we can. That's what we're concentrating on for the moment.

"What if people could actually listen to Christopher Columbus tell his story? That's what we're trying to do. We want to get the story about early aviation from the people who lived it."

Osborne blames the whole business of racing around the country trying to track down these aviation pioneers on Neta Snook Southern.

But she laughs when she says it. "If Amelia had never met Neta, I would never have met Neta and none of this would have happened," she said.

Lt. Col Yvonne C. Pateman - USAF (Ret) Author/WASP
Lt. Colonel Pateman (Ret.) was born in Seawaren, Woodbridge Township, New Jersey. She learned to fly at Wutsboro, New York, between 1942 and 1943. She entered the Army Air Force in March, 1943, and was one of two thousand women accepted for flight training. She earned her wings in the WASPs in September of that year. During World War II she flew as an engineering test pilot and aircraft ferrying pilot. Duty assignments took her to: Avenger Field,

Sweetwater, Texas; Romulus Army Air Field, Romulus, Michigan; and Shaw Army Air Field, Sumter, South Carolina.

Pateman remained in aviation after the WASP program disbanded in 1944. She received her commercial license from the CAA, following service in the WASPs her instructor's rating in May 1950, and her instrument rating in February 1976. She has logged over 5,000 hours.

In the five years between World War II and the Korean War, Colonel Pateman delivered civilian aircraft from factories to airfields throughout the country. She also worked as assistant manager at Monrovia Airport, California, and as chief flight instructor at Culver City Airport, California. See Fig. 9-5.

She accepted a commission in the United States Air Force and was called to active duty in 1951 at Hamilton Air Force Base, California. There she served as an intelligence officer at all levels of Air Force units from squadron to a national agency. Pateman taught in the Air Force Intelligence school for three years. Her overseas assignments took her to the Philippines during the Korean war, Japan in the late 50's, and to Viet Nam where she was Chief of the Warning Division for the 7th Air Force during 1969-70. Colonel Pateman's last assignment prior to retirement in 1971 was as Chief of the China Air and Missile Section of the Defense Intelligence Agency.

During her 20-year Air Force career, Pateman received: the Bronze Star for exemplary service in Viet Nam; the Meritorious Service medal for outstanding performance of duty with the Defense Intelligence Agency; the Joint Service Commendation medal for her exceptional performance with the Alaskan Command; the Outstanding Unit award ribbon with one Oak Leaf cluster; the bronze "V" for valor during her tour in Viet Nam; the Viet Nam Service medal with four battle stars; and the Republic of Viet Nam Gallantry Cross with Palm Unit Citation, for service in the combat zone.)

While on active duty, Pateman completed work towards her Bachelor of Science degree from the University of Maryland, and graduated in 1962, through operation "Bootstrap." She completed the Air Force Squadron Officers School and the Command and Staff College by correspondence. She attended the USAF Intelligence Photo-Radar Officers School at Lowery Air Force Base, the Technical Instructors School at Sheppard AFB, the Nuclear Weapons Orientation Course at White Sands, New Mexico, and the Cartographic Orientation Course at the Aeronautical Chart and Information Center in St. Louis, Mo. During her "off duty" time she was active in the operations of military flying clubs both in the states and overseas.

Following her retirement from the Air Force, Colonel Pateman joined the Military Flying Club at the Marine Corps Air Station, (MCAS) Quantico, Virginia and has enjoyed many cross-country flights throughout the country. She currently flies with the Military Flying Club at El Toro MCAS, California. She has written short stories and articles on aviation for which she has won several awards. Much of her writing has been about women in aviation. She has been published in *Minerva*, a Quarterly Report on Women and the Military, and in

Fig. 9-5 Lt. Col. Yvonne Pateman

Aviation Quarterly. In 1990, she produced a 23-minute video, "We Were WASP" using World War II motion picture film from the Air Force Archives. She has completed a 450 page novel on women in aviation during the exciting decade from 1942-1952 entitled *Tomorrow Comes Last* and is currently seeking a publisher.

Colonel Pateman has spoken to numerous groups such as the DAR, the BPW, the Zonta, the Ninety-Nines, and on university campuses on "Women in the Military" and "Women in Aviation."

Colonel Pateman is a member of the Ninety-Nines, the Veterans of Foreign Wars (VFW), the Disabled American Veterans (DAV), the Virginia Aeronautical Historical Society (VAHS), the Silver Wings Fraternity, the Women's Overseas Service League (WOSL), the Aircraft Owners and Pilots Association (AOPA), and the Women Military Aviators WMA) Inc. where she chairs the National Memorials Committee. She was president of the Women Airforce Service Pilots (WASP) 1990-1992.

Colonel Pateman is a charter member of the Women in Military Service for America Foundation (WIMSA) and serves as a permanent member of the national board.

Doris Lockness - Flight Safety Education

Doris Lockness has been in love with flying for more than 54 years. She has been sharing this love and aviation experiences while promoting flight safety almost as long.

Doris first lifted off into the sky back in 1938 and has been there ever since. After Doris got her private license, she worked at demonstrating light aircraft, illustrating "the ease and safety of flight," to the general public. When World War II broke out, Doris was working as a liaison engineer on C-47s for the Douglas Aircraft Company and she became one of the first women to join Jacqueline Cochran's women's flying unit which later became the WASPs.

Doris' flying experiences have given her a sound foundation for her expertise and devotion to flight safety. As the years literally flew by Doris began amassing an impressive flight log and collection of awards. See Fig. 9-6.

Her personal contributions to the promotion and public acceptance of general aviation have been significant throughout the years. She has also been instrumental in directing countless persons toward flight or aviation careers. Probably Doris' greatest achievement has been her contribution to safety. "Flight safety has always been a first consideration and my proudest achievement," said Doris. "In my 54 years and over 8,000 hours of flight time and giving

Fig. 9-6 Doris Lockness

instruction to pilots, I've never been involved in a flying accident of any type."

In her never-ending quest to promote flight safety and pilot proficiency, Doris is an ardent supporter of the FAA pilot proficiency (Wings) program. Doris not only preaches safety but walks the talk. She is an ardent supporter of pilot recurrent flight training and has personally participated in this program since its inception in 1980. In 1991, Doris received her Phase VII Wings from the FAA.

In 1963, Doris joined an exclusive club, at the time, that numbered only several dozen members. She became the 55th woman in the world to receive her helicopter rating, thus making her eligible for the Whirly-Girls, the International Women Helicopter Pilots. Today that organization numbers more than 900 members. She followed that rating with becoming the second woman in the United States to obtain a commercial type rating in a constant speed propeller-driven gyroplane, in 1988.

As a life member of the OX5 Aviation Pioneers, her early aviation experiences were recognized by this organization. In 1984, they presented her with their "Legion of Merit" award. Doris is the only woman pilot recipient of this award and the organization followed it up with their "Outstanding Women's Award" in 1987.

Again in 1989, Doris was honored by induction into the Aviation Pioneers Hall of Fame, the highest honor awarded.

Doris has also been influencing public acceptance of the safety and reliability of air transportation through her own aeronautical experiences. Doris holds all FAA pilot ratings, she is current and maintains proficiency in several types of airplanes, helicopters, free balloons, and gliders. She also holds flight instructor ratings in all categories of airplanes and hot air balloons.

Doris has received international recognition in recent years through newspaper feature stories, magazine articles and several television interviews.

Doris Lockness currently holds memberships and is active in numerous organizations: Twirly-Birds, Pioneer Helicopter Pilots, Balloon Federation of America, Confederate Air Force, Women Military Aviators, and the Helicopter Club of America. She is the only pilot in the world qualified for membership in all five of the most exclusive flying organizations in the world, the Ninety-Nines, Women's Airforce Service Pilots, Whirly-Girls, OX5 Aviation Pioneers, and the United Flying Octogenarians. In addition to her many duties promoting flight safety, Doris participates in air shows where she flies her restored World War II Vultee-Stinson L-5, affectionately called, "Swamp Angel."

Nancy Hopkins Tier - International Women's Air & Space Museum

Pilot pioneer Nancy Hopkins Tier has witnessed aviation come of age in our country. Her pilot's license was signed by Orville Wright in 1930 and her flying friends have included women like Amelia Earhart and Ruth Nichols. As a charter member of the Ninety-Nines, she was in attendance at the inaugural

Fig. 9-7 Nancy Hopkins Tier

meeting in the little hangar at Curtiss Field in Valley Stream, New York and has dedicated a lifetime of service to their causes.

During the early air race and derby era, Nancy was a sales representative for the Viking Flying Boat Company's "Kittyhawk" airplane. In 1930 she flew their plane more than 7,000 miles as the only woman in the grueling 5,000 mile "Ford Reliability Tour" and in the 2,000 mile "Women's Dixie Derby. "She was the"Connecticut Speed Champion" in 1931, and has left a trail of other trophies and titles to commemorate her racing spirit.

In 1942, Nancy joined the Civil Air Patrol where she served for more than 18 years. She was their first female Wing Commander, and their first female Major, as well as a Lt. Colonel and Colonel and a member of the National Command Advisory staff. She has been honored for her "Aviation Achievements" by the Wings Club in 1983, the C.W. Post University, in 1976, and for her "Outstanding Service" by the Civil Air Patrol in 1981. She is an honorary member of the "38th Strategic Missile Wing: USAF," and of the Aeronautical Medical

Fig. 9-8. Shown at the dedication/anniversary of the IWASM, (l) Nancy Hopkins Tier, Cap. Lori Griffith, Susan Schulhoff-Lau, Bernice Steadman.

Association and was named New England's "Woman of the Year" in 1976. See Fig. 9-7.

Privately, Nancy has owned ten different airplanes and boasts 65 years of continuous flying. She is a member of the OX5 Pioneers, The Early Bird Association, AOPA, EAA, the Silver Wings, The One-Seventy Association, The Aero Club of New England and the UFO, (United Flying Octogenarians.)

Throughout her life, the active aviation enthusiast has supported and sustained the advancement of women in aviation, but her most impassioned fervor has been devoted to the creation of the International Women's Air and Space Museum. Nancy shared the dream of developing a museum solely devoted to aviation's women along with originators Alice Page Shamburger and Bernice Steadman. Through their vision, the world would learn of the impact women have made in the field.

An official board was formed in 1969 under the auspice of the Ninety-Nines and feasibility and impact studies were completed to determine logistics of such an undertaking. A professional research team was hired to conduct a field study to locate the best possible site for the museum and Dayton became their obvious choice. "We looked at half a dozen different places throughout the country but with so much aviation history resting in the Dayton area, it seemed the most logical place for us to be. The Ashael Wright House, (great-uncle to Katharine Wright and her famed brothers), had just been restored in Centerville and the city fathers were looking for an aviation interest to occupy the space. We were delighted to adopt it as our first home."

The museum was incorporated as a non-profit organization in 1976 and dedication cermonies for the "first" and only women's aviation museum were conducted in March 1986. In the course of some seventeen years, many women have lent their time and talents to the cause of seeing the museum dream become a reality but the four most instrumental to its success include: President Nancy Hopkins Tier, Executive Vice President Bernice Steadman, Administrator Joan Hrubec and museum Treasurer Susan Schulhoff Lau. Their aim is to "preserve the history of women in aviation and space and the documentation of their continued contributions today and in the future."

"I don't doubt that we have the greatest collection of women aviation memorabilia in the world," remarks Nancy proudly. Displayed within are significant items from pilot pioneers, astronauts, record setters, WASPS and women from virtually every facet of aviation. The museum also houses an extensive reference library and schedules thought-provoking lecture series throughout the year. Funding is secured by yearly sponsorships and memberships as well as from contributions and donations.

Exhibit expansions, growth of their extensive collections and increased attendance forced the International Women's Air and Space Museum to outgrow the Wright home after just five years. Plans are complete and fund raising has begun for a spacious 30,000 square foot facility that will house aircraft,

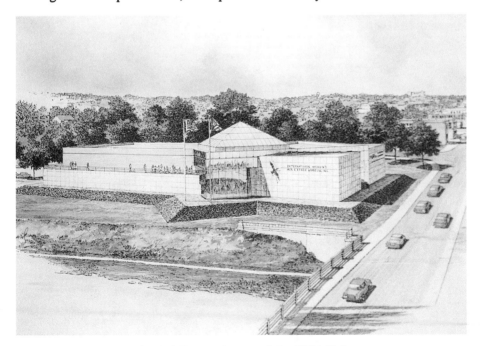

Fig. 9-9 Future home of the IWASM

exhibits, a sensory theater, archives and a research center. See Fig. 9-9 .

Appropriately, it will be located in downtown Dayton, just three blocks from the Wright Brothers' bicycle shop, the spot where the dream of flying first "took off." It will be fully operational for the 100th Anniversary of Powered Flight in the year 2003. .

Nancy Hopkins Tier can be proud of the legacy her efforts have created. The International Women's Air and Space museum will perpetuate an interest in women's aviation by preserving its past and will insure its future by inspiring generations to come.

Chapter Ten

Airline Pilots

*My successes are a source of my joy and my failures are a source of my wisdom.
I can, if I try.* Connie Tobias

Helen Richey

Helen Richey earned her private license in 1930, at the age of 20, and her father celebrated with her by giving her an open cockpit biplane. Then she announced she was going to become a commercial pilot for an airline. She realized that an airline job was in the future, so she began building time as an aerobatic pilot. That led to a job with Curtiss Wright. She finished third in the Amelia Earhart Trophy Race in 1932, and earned national popularity. In 1933, she was the co-pilot for Frances Marsalis when they set a 10-day endurance record. When the hose of the refueling plane ripped the fabric of their plane, Richey climbed out on the fuselage behind the wings and mended the torn fabric. See Fig. 10-1.

In 1934, Helen Richey applied to several airlines for a pilot's job. Coincidently, Central Airlines and Pennsylvania Airlines were in stiff competition for the same mail routes, and Central Airlines realized there was great publicity value and novelty if they hired a woman as a pilot. In 1934, Central Airlines broke precedent and hired Richey. Helen Richey became the first woman to pilot a commercial airliner on a regular scheduled route. She made her first flight as a pilot on December 31, 1934, flying a Ford Tri-Motor from Washington to Detroit. The newspapers prematurely hailed the move as breaking new ground for women in aviation and, "the dawn of women coming of age."

Soon Richey became increasingly aware that the company was using her more as a public relations agent than as a pilot. She found herself on the lecture

Fig. 10-1 Helen Richey

circuit giving interviews, posing for pictures, handing out autographed souvenir postcards to school children, and not doing much flying. There was also vehement opposition and blatant discrimination from the rest of Central Airlines' pilots. The male pilots rejected her application for union membership, and because she was a woman, the CAA warned Central Airlines not to let her fly in bad weather. After 10 months she resigned from her job, saying she was not going to be a fair-weather pilot. Central Airlines had let her fly only about a dozen round trips.

The WASPs were highly qualified to fly the airline equipment in 1946, when civilian aviation restarted after World War II, but were not given the opportunity.

The first step toward progress in this area of aviation came in January 1973, when Emily Warner became first officer on a Convair 580 for Frontier Airlines. The door had been opened ever so slightly on the last sex-segregated aviation occupation in the non-military aviation industry.

Warner had been flying since the age of 17, and she had held jobs as diverse as flying traffic reporter, FAA examiner, and flight school manager. She was the chief pilot for the Clinton Aviation Company in Denver when she first applied to Frontier in 1968. The employment opportunities for pilots were limited, but when Frontier decided to recruit a new group in 1972, Warner applied. She was persistent, and when they finally hired her, a spokesperson for the airline said, "We couldn't think of any reason not to."

Two months after Warner was hired, Bonnie Tiburzi became the second woman to fly for a scheduled airline, American Airlines, and she is considered the first for a major airline (Frontier, at the time they hired Warner, had only regional revenue status.)

Tiburzi came from an aviation family. Her father was a pilot during World War II with the Air Transport Command for TWA. He took her on her first ride when she was twelve. By the age of twenty, she was flying co-pilot for charter airlines in Europe. Her brother Allen was the youngest pilot to fly the largest transport aircraft at the time, the first stretch DC-8, for Seaboard World Airlines. He became a captain on that aircraft at the age of 31, in 1978. Bonnie's grandfather was also the first man in Sweden to manufacture aircraft parts.

The obvious question is: "why weren't women hired before 1973?" The airlines claimed that qualified women had not expressed interest or applied. Economic and marketing priorities were more decisive factors, however. Highly regulated by the federal government and unable to manipulate either routes or fares, the airlines competed for passengers through service. Image was extremely important. In the early 1970s, there were still significant numbers of Americans who had never traveled, or traveled only once on an airliner. Even for trips over 200 miles, most Americans still drove their cars. Consequently, the airlines were still very much wedded to a public relations program of safety and convenience. They deliberately cultivated the image of the pilot as "father" and were uncertain as to whether or not putting women in the cockpit might instill new fears in passengers' minds.

The airlines had little experience with women as career professionals and women were not seen as primary wage earners: men needed to support families, women did not, or so the prevailing perception. The argument Jacqueline Cochran had advocated in 1933, was valid in their eyes 40 years later. The airlines argued that women would soon quit to start a family, thus supposedly validating the notion that hiring women would not be cost effective.

In the 1970s, the airlines had to become "Equal Opportunity Employers" if they were to receive federal money. Further, the equal rights activists lobbied the companies. The movement had much to gain if women were hired. Airline pilots were highly visible and well regarded. The National Organization for Women wanted visible role models of women in non-traditional careers to demonstrate the legitimacy of its philosophy and goals.

Regulations and lobbying aside, the airlines did not have many women knocking on their doors. The requirements were stiff. An applicant had to be at least 23 years old; have four years of college; a commercial license; an instrument rating; an airline transport rating; 1,500 hours flying time, including both night and cross-country flying experience; and a first-class medical certificate. In addition, knowledge of regulations, and the principles of safe flight were needed. They were required to have exceptional hearing as well as sight and meet physical requirements such as minimum height (five feet, six inches for American; five feet, eight inches for Delta). But the requirements were not the problem - women could and did pass the tests. The problem was attitudes and economics. And attitudes can sometimes take years to change.

These women were subjected to a series of policy rules that prevented them from wearing make-up, having pockets on their flight jackets, or making calls over the public address system of the aircraft. Also, they were generally urged to keep their hair short or tucked in under their cap. All of these things were to hide the fact from the passengers that there was indeed a woman in the cockpit. By the time Susan Maule came of age, attitudes and perceptions were changing.

Susan Maule

Being a "Maule" there was no doubt that Susan would learn to fly, it was just a question of when. As the granddaughter of B.D.Maule, the aircraft designer, flying was in her blood and became second nature. With her father as her flight instructor, Susan's lessons began at the early age of seven. "I wasn't big enough to see over the instrument panel so my dad rigged a special seat for me so I could see the gauges. I couldn't use normal visual references so I had to learn how to fly by the instruments."

The Maule aircraft originally came off the production line when Susan was just two years old. The versatile four-place taildragger has the ability to take off and land within a short 1,500 feet. It is a popular model for bush pilots and grass strip enthusiasts but not the type airplane that most people learn to fly in today.

Susan was set to solo on her 16th birthday and it seemed that date would not come soon enough. October 19th brought a dark and dreary day but it didn't dampen Susan's spirits. "I was up early and worried about the weather. I'd been looking forward to soloing for so long and I wanted to get going. When I couldn't wait any longer I nudged my dad out of bed so we could head to the field." Maule field in Napoleon, Michigan was the original site of the Maule aircraft factory and the spot where Susan had done all her flying.

Not only did Susan solo that day in the taildragger her grandfather designed and built, but she soloed 12 different airplanes, making three take offs and landings in each. "My dad wanted me to solo first in a Maule so we got the M4-220 model out of the hangar. From that point on, friends who knew it was my birthday kept showing up and flying in to let me fly their airplanes."

The rest of the 12 airplanes Susan soloed included a 1939 Aeronca Chief (taildragger), Cessna 150, Piper Cub, Aeronca Champ, Tri-Pacer, Cessna 172, Warrior, Aero, Taylorcraft and a Maule M4-145 model. A trip out to Wolf Lake even added a seaplane solo to the day as she made three take offs and landings in a Citabria on floats. Susan quit just in time to make it to the Department of Motor Vehicles so she could get her drivers license before they closed.

"In order to solo on my birthday I had to skip school but I sure didn't get away with it. The next day the news about my day was spread all over the front page of the local paper and the whole town knew that I'd played hooky."

Fig. 10-2 Susan Maule

Susan's enthusiastic pace continued and it was no surprise that she received her private ratings, land and sea, on her 17th birthday. "Growing up with flying in my back yard made me think that grass strips and the great older model airplanes were all I needed. I didn't know any airline pilots and didn't even consider turning my love for flying into a career."

Susan attended Valdosta State College and graduated with a history major. Completing the student teacher phase made her realize that her chosen field wasn't as exciting as airplanes, so she went back to flying. Like clockwork she earned her ratings and received an Amelia Earhart Scholarship from the Ninety-Nines for her ATP. See Fig. 10-2.

To earn money Susan worked in the Maule factory in the covering department, stretching fabric over elevators, rudders and stabilizers. To build flight time she ferried Maules across the country to waiting owners and gave demo flights to would-be buyers. Flight instructing in San Diego allowed her the flexibility to be the west coast representative for her grandfather's company.

In 1985, Wings West, a commuter airline in California, hired Susan. Within nine months she upgraded from First Officer to Captain on the Metroliner. Susan's airline dream came true when Piedmont Airlines hired her in 1986. The Fokker FK-28 was her first aircraft assignment and she was based in Syracuse,

NY. Susan upgraded to co-pilot on the B-737-300/400 and flies today out of Baltimore, Maryland.

Flying taildragger models requires a unique skill that must be learned. One needs a special knack or touch combined with a sharpened sense of awareness. Susan learned this early in her flight training and is what has made it so easy for her to transition to so many aircraft types. With more than 100 different kinds of airplanes tallied in Susan's logbook, she is quick to point out her favorites. "There's no doubt that the B-737 and the Maule are my favorites, they are the perfect balance. The hi-tech wizardry of the new 737 models puts me on the cutting edge of aviation today. The tried and true Maule is a thrill to fly and takes me back to the golden days which is what made me first fall in love with flying in the first place."

Today Susan is active in the Ninety-Nines as well as the International Society of Women Airline Pilots. She enjoys speaking to youth groups about aviation and has participated in three Air Race Classics, all in the airplane that bears her name.

Karen Kahn

Karen Kahn took a $5.00 introductory flying lesson in 1968 and was hooked. "They let me take the controls on that first lesson and I've had a tough time letting go ever since." She scraped enough money together to cover the cost of flying and scheduled lessons when she could afford them. It proved to be a time consuming approach to completing her ratings but Karen was persistent. "Admittedly it would have been quicker and cheaper if I'd saved all the money the rating required prior to beginning my flight training but patience was never one of my virtues."

Karen's parents were ambivalent about her new found love for flying. Letters from home often contained newspaper clippings detailing the most recent airplane crash but it wouldn't shake her fascination. Karen completed all her ratings and built her flight time as an instructor. She took a job in the Midwest with a weekend ground school company, but the company declared bankruptcy two months after they hired her.

Karen and another instructor studied the mistakes that the previous management had made and decided to revive the company and make a go of it on their own. Their new company, Accelerated Pilot Training, was born in 1975. "The odds were certainly against us making it but I saw it as an opportunity to build flight time by owning a business that required me to fly. I flew to a different city each weekend teaching pilots how to pass the FAA written exams for their private, commercial and instrument ratings." Geographically, most of the ground school business, as well as the student pilots, were centered in the California area so they relocated the company. In just two years, the little business Karen helped to build was really flying. When her partner offered to "buy her out" she seriously studied her options. "At that point I had 2,000 total flight hours and an ATP certificate in my pocket. This was the perfect chance for me to take the money and run and start searching for the airline job I'd always dreamed of." See Fig. 10-3

The year was 1977 and although there were a handful of women who had already been hired by major airlines, Karen left nothing to chance. She requested applications from every major carrier and disguised her gender by using only the initials of her first and middle names and marked "N/A" in the squares that asked for military experience.

It was a friend that suggested she consider a smaller, growing carrier like Continental, so she filled out an application and submitted it in person. Karen was promptly called to undergo the interview process and was invited to join the new-hire pilot class on July 11, 1977. "It was the job of a lifetime. I had reached my goal and intended to work my brains off and be a success at it."

Karen's career started as a second officer at the flight engineer's panel of the B-727. Although she was the fourth woman to fly within Continental's ranks, she frequently came across crews who had never seen a woman pilot let alone flown with one."Most of the people I flew with were friendly and glad to see a woman in the cockpit, or so they led me to believe. Those that weren't openly enthusiastic were at least fair enough to remain neutral. They preferred to wait and see how I did my job before making up their own minds. In either case I was very proud to be working with these men."

A positive attitude coupled with a genuine interest in learning was Karen's key to successful survival in those early years. "I just let them know that I didn't pretend to know everything and appreciated any of their suggestions. I made it clear that I didn't consider myself to be anything special, just part of the crew, trying to do a good job."

Karen upgraded as a first officer but experienced the "yo-yo" effect as the airline's first surge of furloughs relegated her to reserve flight status. She opted for a change of scenery both in and out of the cockpit, and bid out to the western Pacific as an international second officer flying for Continental's subsidiary, Air Micronesia.

Being based in Honolulu and flying to far away places with strange sounding names was not as glamorous as it appeared. "A typical day at 'Air Mike' had us departing Honolulu to begin a 14 hour duty day. We flew 10 hours of flight time, landed at six different islands with minimal to no navigational facilities, in the worst of tropical weather situations all while crossing four time zones and the international date line." Karen flew two six-month tours for Air Micronesia, an experience she will never forget.

As a female airline pilot pioneer flying in a male-dominated environment, Karen wondered how the first women at the other carriers were faring. She was curious to learn if they had encountered some of the same challenges and longed to share her ideas and experiences with them. This interest grew into a position as a founding member of the International Society of Women Airline Pilots (ISA + 21), and she was the organization's third president from 1982 -1984. While the group was formed as a simple social affiliation in the seventies, Karen

Fig. 10-3 Karen Kahn

helped professionalize and expand its scope to meet the needs of the growing women airline pilot population of the eighties. In 1983, Continental declared bankruptcy and every airline employee was laid off. Days later the management re-started operations with a skeletal force and a general strike ensued. Experiencing a furlough and a strike were painful reminders that no one was immune to unemployment in the vulnerable airline industry.

Karen managed to find a job in the corporate world, flying for Global Airways and became the first woman in history to earn a Lockheed Jetstar type rating. "Corporate flying was worlds apart from the well-organized, tightly-scheduled flying I'd done for the airlines. It took some adjustment but I was grateful for the job and the chance to expand my flight experience while Continental settled its internal problems."

Karen upgraded as a captain on the MD-80 with Continental Airlines, in May of 1988 and continues to enjoy the view from the left seat. "I still greet all my passengers over the PA on the ground before each flight. I figure if anyone is uneasy with the thought of a woman behind the controls they have plenty of time to leave. So far no one has ever walked off."

Today Karen does more than just fly the big jets for a living. She has developed a unique aviation counseling service that helps to unravel the confusing options that face aspiring pilots today. In the current aviation market, pilots compete for a limited number of openings and Karen believes that effective counseling can save precious time and money, ultimately improving their employment opportunities. With twenty two years of aviation knowledge and experience, she is well qualified to put serious pilots on a track that can lead to a professional pilot career.

"I realized that I'd been giving basic career counselling for years at the many pilot seminars and conferences where I was asked to speak. By designing an actual business, I could reach more people and devote more time and energy to pilots and their aviation goals."

Sessions have been created to include, introductory, professional pilot and airline interview counseling. Karen and her staff make themselves available to answer questions, make recommendations and access alternatives based on the individual's total flight time, education and finances. She is also prepared to assist established pilots to research medical problems and airman certificate violations.

Karen provides assistance at every level, answering questions like, how to enter the field, what flight school to attend, how to finance training and financial aid options available. Those building flight time wonder about getting the most flight experience for their money and investing in advanced ratings. Even learning how to be more marketable and how to improve interview skills are areas where competent advice can be invaluable.

"Pilots need to have an aggressive plan-of-attack where their careers are concerned. We help pilots present themselves and their qualifications in the best light possible." Those who have taken advantage of the service have found it encouraging and motivating to get support from a pilot who remembers the trials and tribulations of getting started. A single statement found on the program's brochure serves as the biggest inspiration for the pilots Karen helps to guide, "Average People Consider The Sky The Limit. Pilots Think Of It As Home!"

Karen Davies Lee

Karen Davies Lee started flying in 1968, before her 17th birthday. She earned her ratings and flew a typical general aviation profile which included flight instructing, charter, towing gliders, flying C-46 freighters in the Caribbean and auto parts charters out of Michigan. She progressed into the corporate world of flying before becoming the first female in history to fly for TWA.

On October 2, 1978, Karen earned the distinction of being a "Female Airline Pilot Pioneer" as another of the major carriers opened its doors to a woman. "Someone had to be the first," remarks Karen. "I was just in the right place at the right time with the right qualifications." She completed ground school, survived her probationary period and was just settling into her position as a B-727 flight engineer when she received her furlough notice. Fortunately,

Fig. 10-4 Karen Davies Lee

Seaboard World Airways hired Karen almost immediately to fly as a DC-8 First Officer. She was assigned to their base in London and flew scheduled cargo service on contract for Saudi Arabian Airlines. See Fig. 10-4.

In the fall of 1980, Karen was furloughed again. First, for just a few months after the merger between Seaboard and Flying Tigers, and again in 1981 when the contract with Saudia Arabian Airlines was not renewed. It was a struggle to find a flying job because the market was flooded with qualified pilots who were furloughed throughout the industry. She launched an exhaustive search that lasted nine months before she was finally able to find a job with Orion Air. Orion was a small supplemental carrier that converted G-1's to freighters and operated them for UPS and Emery Air Freight. They had just won a contract to fly eight B-727's for UPS and Karen was quickly assigned as a B-727 flight engineer.

"During my long hiatus of unemployment I desperately looked for another flying job or a position outside the aviation industry. What I learned was that my flying credentials weren't worth much to the outside world. Even my degree in aeronautical science did little to enhance my resume. The previous ten years of airline instability had caused eleven different moves, twelve different flying jobs, three furloughs, and months of unemployment that left personal scars that would take years to overcome. I vowed that I would work towards broadening my experience beyond the cockpit at my first opportunity.

"When I interviewed with Orion I practically begged them to let me do something in addition to flying." Orion obliged and besides flying, Karen worked in the flight department screening resumes and filing applications. Within three years Karen progressed both in and out of the cockpit. She had upgraded as a captain on the B-727 and was promoted as Manager of Flight Crew Planning and Administration. "I discovered that I did have abilities beyond the cockpit and began to regain my confidence and self-esteem."

In mid-1985, nearly seven years after Karen had originally been hired by TWA, they issued her an official recall notice to return. With changes occurring within Orion's management, she decided to accept. Remaining true to the vow she had made to herself, she joined the training department at TWA so she could learn another aspect of the airline industry. For the remainder of her career there she instructed second officers on the B-727 and B-747 flight engineer panels.

United Parcel Service announced the start-up of their own airline in August of 1987. A few pilots that Karen had worked with at Orion were immediately hired into management positions and when they learned that UPS was looking for qualified women to join their management team, they called Karen. "At the time I was five months pregnant. It didn't seem likely that a brand new airline would be interested in hiring an employee that was destined for an extended leave of absence. They reminded me at my interview that being pregnant was a condition not a disability, and I was promptly hired as a B-727 Fleet Supervisor."

Karen spent the next 4 months working in the personnel department with the task of hiring pilots for the new flight operation. Her previous experience with pilot hiring was exactly what they needed to set up selection guidelines and get the employment process started.

When she returned from maternity leave she assumed her duties as a Fleet Supervisor. "UPS believes that every employee, including the pilots, should know their supervisor on a more personal level. This is a more progressive concept than most traditional airlines adopt." With 35 pilots under her direct supervision, Karen acted as their vital link and attended to all her pilots' administrative needs and operational concerns.

In addition to her supervisory role, she instructed the newly up-graded and transitioning pilots in their line duties and gave annual checkrides. In an engineering capacity, Karen developed the performance chapter of the B-727 aircraft operating manual and produced take off and landing speed charts that eventually became standard in every UPS aircraft.

After Karen's second child was born in May of 1991, UPS promoted her as their Next Day Air Flight Operations Manager. This highly visible position held Karen directly responsible for the coordination of all of UPS' flight operations. "My team was devoted to responding to any problems that arose during the night's operation in the flight department. We coordinated with other operating departments and assisted to ensure that the planes and packages would arrive into our hub safely and on time and ultimately get back out to the customers." Massive coordinating efforts such as this confirms why UPS runs "the tightest ship in the shipping industry."

Today Karen has progressed to a position in the airline industry that few, if any, women have ever managed. She is UPS' B-727 Fleet Manager, (the equivalent of B-727 Chief Pilot.) She oversees the company's 310 B-727 flight crew members, a staff of 10 Fleet Supervisors and 12 of the aircraft's line training instructors. Virtually every aspect of operations pertaining to the B-727 both in the air and on the ground, rests on Karen's shoulders. "I am delighted with the opportunities of this new job. The biggest challenge I face is providing quality leadership to an exceptionally well qualified group of supervisors and instructors, any number of whom could be my "boss" if the timing had been different."

On the horizon Karen will be responsible for the implementation of UPS' Quiet Freighter Program. This multi-million dollar project has been slated to totally re-configure the entire B-727 fleet to quieter and more fuel efficient engines by 1994. Karen has proved that hard work and determination can overcome adversity. Behind Karen's personal and professional success stands many people in her life and in the UPS family that have encouraged her, believed in her and have opened doors that have allowed her the chance to prove herself. They are the motivating force behind Karen and thousands of other UPS employees who take pride in their work to bring us the quality service we enjoy everyday.

Connie Tobias

A question often asked of female airline pilots is, "When did you decide that you wanted to fly airplanes?" For Connie Tobias, currently USAir's only female B-727 captain, the answer is: "Ever since I can remember! I always felt a yearning to fly even though society opposed the idea of women as pilots when I was growing up."

Connie refused to allow society's limiting attitudes to discourage her interest in aviation. She had the motivation and the determination required to become a pilot but she lacked the financial ability to afford the expensive quest. Connie tried replacing her desire to fly with other adventures hoping to fill the void. She co-founded a wilderness excursion group and explored new horizons. In just three years with the group she experienced more excitement and adventure than most people do in a lifetime. She sailed the Caribbean and Atlantic Ocean, became SCUBA certified, rode horses, bicycles and motorcycles, climbed mountains and rafted the fastest navigable waters in the world.

It was during a 3,000 mile 10-speed bicycle trip across America that Connie stopped at a bridge in the Midwest and saw something that would give her life new direction. "I stopped to get a drink of water just as a jet airplane took off and climbed out overhead. I watched until it disappeared into the clouds and finally realized that what I really wanted to do with my life was fly. I decided at

Fig. 10-5 Connie Tobias

the completion of the bike trip I would go home to Ohio, sell everything I owned and become a pilot. I figured it would be better to shoot for the moon and miss than never take aim at my goal of becoming an airline pilot." See Fig. 10-5.

Connie attended Ohio University and put herself through college with the aid of scholarships, grants and loans. She worked simultaneous part-time jobs to pay for her flight training and endured many hardships in her struggle to make ends meet. "Many a hot meal came in the form of peanut butter on toast but the joy of flying was in my blood and I was determined to make whatever sacrifices were necessary to follow my dream."

Connie was a serious and conscientious student whose academic achievements earned her a Presidential cabinet position at the University. She served as President of the National Honorary Mortar Board and was selected as "Aviation's Outstanding Student of the Year" in 1977. She graduated from Ohio University summa cum laude with two aviation degrees and an academic Masters in Engineering.

Progressing through the ranks of civilian aviation, Connie built her flight time by instructing, flying fire patrol and charter. She flew for commuter and regional airlines before finally being hired by Piedmont Airlines on May 7, 1984.

"Making it to the majors made ten years of hard work and sacrifice feel like an overnight success."

Piedmont flourished following airline deregulation and initiated a surge of growth when Connie came aboard. Rather than the usual entry level flight engineer position, she began her airline career as a B-727 first officer and upgraded to captain on the Fokker F-28 just two and a half years after being hired. She upgraded as a captain of the B-737-300 in 1989 and again in 1990 as a B-727 captain.

Today Connie flies over that same bridge in the Midwest where she once stood and looked up with the dream to fly. "My job is challenging and full of rewards. I'm in the business because I love to fly not because I'm trying to make a statement." The fact that she and so many other professional women pilots have made a positive impact in airline cockpits all over the world serves as statement enough.

Connie encountered individuals along the way who questioned a woman's presence in the cockpit, but for every person who was negative there were scores of others who provided positive support and encouragement. "It takes a man of quality to accept a woman of equality. I thank those men of quality who accepted me and reinforced my quest to fly. I thank those courageous women who have so kindly guided me and who have been my friends. I must also thank those who challenged my ability for in part, their doubt and opposition made me stronger and drove me to become an even better pilot."

A love of flying, confidence, commitment and minimal use of the word "can't" is Connie's advice for aspiring pilots. Still taped to her roll top desk is a quote that has been inspirational to Connie in her aviation pursuit, "My successes are a source of my joy and my failures are a source of my wisdom. I can, if I try."

Connie is a member of ISA + 21 and was inducted into their Captains Club in 1986. She is on the Aviation Advisory Board for the Department of Aviation, College of Engineering at Ohio University and also serves on their National Alumni Board of Directors.

Besides regular line flying, Connie devotes her time and energy to a Crew Resource Management (CRM) program which she helped to develop at USAir. More than just an airline buzz word for the 1990's, CRM training has been introduced at virtually all levels of aviation. The concept strives to improve crewmembers' operational effectiveness and teaches them how to maximize all available resources to resolve conflicts that may ultimately threaten a flight's safety.

Connie has been dedicated to the project from its inception and not only helped design the current curriculum but conducts the course as a classroom facilitator. "We created each phase of the training specifically for our pilot group and their operational environment. The first phase of training serves to introduce the pilots to the basic concept and philosophy of CRM and incorporates hands-on exercises that encourage pilot participation." Connie uses an effective teaching method derived from the adage, "Tell me and I will forget, show me and I will remember, involve me and I will learn."

Actual cockpit flight data recordings and accident films are also incorporated into the training to illustrate the crucial need for implementation of proper CRM skills and techniques. CRM success at USAir is due in part to the vision of those like Connie who are dedicated to improving the airline's safety environment by educating the pilot workforce. "I'm proud to be a part of this timely and innovative program and am confident that its rewards will be immeasurable."

Airline management is an area that also interests Connie and she believes that one day opportunities for women at the management level will be available for those who are interested. "Women have yet to be incorporated into the flight department management team at USAir, but it's just a matter of time before they trade their skepticism for acceptance and realize that the same professional leadership qualities we have demonstrated for years in the cockpit would be a valued asset."

Terry London Rinehart

As the daughter of two military pilots, an Air Force pilot/retired colonel and the only WASP to earn the Air Medal during World War II, Terry London was bound to grow up with an interest in aviation. "I practically grew up flying. We went everywhere in airplanes and even listened to tower and ATC frequencies at home instead of the radio. During the years that my mother was the executive director of the Powder Puff Derby, our whole family became involved in the races. After I got my private license my mother and I even flew as a mother-daughter flying team in the All-Women's Transcontinental Air Race from Seattle to Clearwater, Florida. That was a real treat for me."

Terry learned to fly like most pilots, in a Cessna 150 and soloed on her sixteenth birthday. She built her flight time and experience working as a flight instructor, a ferry pilot, a flight school manager and a dispatcher. In 1967, Terry won the Godfrey Claussen Award as the Most Outstanding Young Female Pilot at the First International Exposition of Flight in Las Vegas, Nevada. She followed that honor by winning the Doris Mullen Memorial Helicopter Scholarship awarded by the Whirly-Girls, in 1969, which she used to obtain her commercial helicopter rating. See Fig. 10-6.

Terry attended California State University in San Jose, CA. She earned her degree in aeronautical operations and graduated with departmental honors. Her advanced aeronautical studies include aeronautical engineering at UCLA, aerospace at Weber State College and aeronautics at Long Beach City College.

Becoming an airline pilot was Terry's real dream. This drive and ambition, coupled with her aviation achievements, earned her the California Aerospace Education Association Scholarship to attend Boeing 727 Flight Engineer School at the Sierra Academy of Aeronautics in Oakland, CA. "Looking back I realize that the idea of becoming an airline pilot seemed unrealistic but the whole time

Fig. 10-6 Terry London Rinehart

I was working towards my goal my mother never told me that there weren't any women flying as airline pilots. She encouraged me every step of the way and by doing so taught me to never set limits on my goals."

Terry landed a job as a jet engine service coordinator for United Airlines and faithfully applied to the major airlines in hopes of pilot employment. "When I showed up at Western Airlines' headquarters building to personally submit my application, the secretary thought it was a joke and to go along with the hoax she sent me up to the Vice President of Flight Operations. Once inside the VP's office he saw my qualifications and realized it was no joke and asked, "So you really want a job?"

It was the first time that the airline was forced to deal with a female pilot applicant and they treated the issue very seriously. They directed Terry to the Chief Pilot's office and although the airlines were at a virtual standstill on hiring due to the country's fuel crisis, he took a noted interest in her progress. "He told me to keep him updated as I increased my flight experience so I submitted a new resume after each advancement." When Terry notified Western that she had received her ATP, she was invited back for a formal interview. She successfully completed the interview process and the pilot applicant physical and on March 8, 1976 was hired as Western Airline's first female airline pilot. She was assigned as a B-737 second officer and both passengers and crews welcomed her presence in the cockpit. "The flight attendants were great promoters. I think they were glad to see a woman make it to the cockpit. They were so positive that the passengers couldn't help but be the same." '

Measuring exactly six feet, Terry is taller than most the men she flies with and admits that she has not encountered any discrimination as a woman in the cockpit during her seventeen years with Western, (now Delta.) Jokingly she quips: "Maybe it's because I'm bigger than they are! Actually the airline industry is probably the most equal opportunity orientated industry there is. They acknowledge our ability and recognize that females can also have a career and a family."

Terry literally met her husband Bob Rinehart, a pilot for United Airlines, on the runway in San Jose when both were participating as judges at a NIFA air meet in 1975. They fell in love and married. Over the years they have mastered the art of juggling two separate airline careers with the hectic schedules and have successfully met the challenges of their three active children.

Terry is a member of the Ninety-Nines, The Whirly-Girls, (International Association of Women Helicopter Pilots) and is also a charter member of ISA + 21, The International Society of Women Airline Pilots.

Finding the perfect balance between her flying and her family life, Terry proves that with compromise and sacrifice it is possible to "have it all." For now she has traded captain's stripes for apron strings and has passed up numerous opportunities to upgrade in order to maintain a seniority that allows her family as normal a lifestyle as possible. "It's a matter of establishing priorities. I love flying, it's what I've been doing all my life and I don't plan to quit until they make me retire at age 60, but flying is only what I do 80 hours a month. The rest of the time I am a full-time wife and mother and that is what's most important." The Rinehart children follow the legendary lead of their illustrious pilot parents and grandparents and it seems certain that a third generation of flyers will emerge. "Children today grow up with the knowledge that they can be anything they want to be," says Terry. "The key is choosing something they'll enjoy doing for the rest of their lives regardless of the occupation."

Cyd Laurie Fougner

Cyd Fougner grew up in Minot, North Dakota, where she got her first job in aviation. She was 15 at the time and worked at the Minot International Airport as a secretary/bookkeeper and receptionist. Eventually she worked her way up to the flight school dispatcher's job. Her first plane ride came at the Makote, North Dakota Threshing Harvest Show, where she sold penny-a-pound rides. That first ride was enough to lure Cyd into the sky and a career in aviation.

Fig. 10-7 Cyd Laurie Fougner

Cyd graduated from high school in 1970, and one month later had her private pilot's license. By the time she was 18, Cyd was working for Aero Flight in Minot, and on her way to a career in aviation. She had earned her commercial license and her flight instructor's rating. It was then time for her to reverse the cash flow. She started teaching students to fly.

By 1972, Cyd began to feel growing pains. It was time again for her to move up and this time move on. She moved to Marin County, California, and enrolled in college. At the same time she was broadening her education, she was expanding her flight experience. In addition to instructing and flying charters, Cyd supported herself by becoming a flying traffic reporter for a local radio station, spreading ashes for burial flights, flying photo missions for local photographers, and adding ground school instructing to her aviation repertoire.

In January 1975, Cyd broadened her horizon again by transferring to San Francisco State University. Just a year earlier she had earned her instrument rating. Cyd continued to earn her living flying and by 1977, had also earned her bachelor's degree. Along the way she had heard about the Whirly-Girls and won their scholarship which she used for her commercial helicopter rating. By 1977,

Cyd had her multi-engine rating and was now ready for her dream job, that of an airline pilot. See Fig. 10-7.

We sometimes live for our dreams and reality sometimes discourages us from striving to make our dreams come true. This was not the case for Cyd, although she discovered that filing an application and a resume did not make an airline pilots job happen. For the next year and a half she filed dozens of applications and got dozens of rejections. The rejections did not stop her either. Nor did the fact that reality dictated she'd need more than a few thousand hours in her log book and a college degree.

Cyd realized that there were thousands of applicants shooting for the same few jobs, with more hours and more qualifications. So what did she do? Cyd went back to school to earn her flight engineer's turbo jet rating in a B-727. She earned that rating in May 1978, and a month later she had her ATP.

Now she knew her dream was within her grasp. She updated all her applications and got calls from nine of the major airlines. Then came the interview process, the tests, physical exams, psychological tests, simulator checks and of course some rejections. But there was also something new in the equation - a dilemma. Cyd Fougner had received four offers! After careful deliberation, Cyd started training with Western Airlines on December 18, 1978. After her training period she began flying as a second officer on a B-737 - the 10th woman hired by Western Airlines.

The next two years were great for Cyd, but there was also the ever-looming reality of the airline industry. Economic cutbacks caused the company to furlough Cyd and many other pilots.

The furlough lasted five years and Cyd used the time to grow both personally and professionally.

During the five year furlough from Western Airlines, Cyd returned to flight instructing and completed her instrument instructor rating and her multi-engine instructor rating. In 1981, she passed her air taxi checkride and began flying charters and corporate aircraft. To further broaden her aviation knowledge, Cyd sold aircraft, but with the depressed aviation industry, commissions were small and unreliable.

In June of 1982, Cyd landed a job with the FAA as an Air Traffic Assistant, and worked with the air traffic controllers at the Oakland Air Route Traffic Control Center (ARTCC). Still looking to broaden her horizons, Cyd transferred to the Seattle ARTCC, where she began working in the training department, training newly hired air traffic controllers. She performed both the pilot and computer operator functions in the simulators. Her instructor background helped her qualify for a promotion to "Education Specialist." She spent the next two years in this position.

Cyd restarted her airline career on November 18, 1985 when Western Airlines recalled her. She immediately began working as a flight engineer on a B-727, based in Seattle, Washington. In June 1991, Cyd upgraded to first officer on the B-727, with Delta Air Lines (who bought Western Airlines).

Cyd is thrilled to be working for one of the strongest and most respected airlines in the world and is also proud to be an ALPA member in this deregulated environment. Cyd has encountered many struggles centered around being a woman in a man's world, but she has overcome every one with 100 percent success. She reports that the adage, "A woman must work twice as hard as a man to prove that she is half as good," is true, however, the work coupled with a GREAT attitude will win over each male crewmember one at a time. "While at work I am a pilot first, and a woman second, but then off duty I am a woman first. "Gender should not be an issue in the cockpit, but in 1992, she accepts her status as "just one of the guys" as a step in the right direction. "I hope for a future," she says, "where occupations like pilot, teacher, nurse, doctor or president do not carry a gender stereotype with them."

Cyd accepts the reality that discrimination and sexual harassment are still common, but she believes much of it arises from naivete instead of malicious intent. Cyd's approach is to diagnose the intent, letting the ignorant or unen-lightened people slide, while dealing with the malicious comments or actions very directly but delicately. An example of just one way to handle the offending person is to look them directly in the eye, and ask them to repeat their comment so that other people will hear them too by saying, "What did you say?" Then after they comply say, "That's what I thought you said." Then turn around and leave the scene. Her objective is to educate them by highlighting their lack of aware-ness that each gender has comparable worth. Cyd's personal goal is to avoid getting a chip on her shoulder. She knows that if someone has a problem with a woman pilot, that it's THEIR problem, not hers. Her job is to fly safely, a job she has been done successfully.

Cyd feels that her job as an airline pilot is the best job in the world for her and that her professional satisfaction has resulted in her wonderful quality and contentment of home life. She is married to Peter Dolliver and is deeply rooted in her community in Gig Harbor, Washington. Although Peter's only foray into the field of aviation is as a paraglider pilot, he fully supports his wife's career decisions. They enjoy hiking, camping, dancing, music and yoga together, while Cyd is the gardner in the family. Cyd is also a certified SCUBA diver and prefers warm tropical waters for her vacation destinations. She is still a current CFI and enjoys teaching the occasional student between airline trips. "I still enjoy motivating a new student with my enthusiasm and passion for the greatest high in the world."

Betsy Carroll Smith

Betsy Carroll Smith started flying in 1968 on her sixteenth birthday. "Al-though I come from a family of artists, we've always shared an interest in exploring the natural world. An uncle of mine, who had lived in Alaska for several years, often told stories about bush pilots he had known and the thought of learning to fly captured my imagination." Her first lesson was a gift from her uncle and it was a inspiring experience. "I couldn't even drive a car yet so you

Fig. 10-8 Betsy Carroll Smith

can imagine how thrilled I was at being able to fly a plane." Betsy paid for subsequent lessons by stuffing envelopes after school. The part-time job paid just enough to cover the cost of one lesson per week at the local airport in Niagara Falls. See Fig. 10-8.

"As a teenage pilot I had no intention of flying for an airline, but wanted to pursue something more 'interesting' such as environmental research or perhaps even the space program as I was an avid science fiction reader and was completely enthralled by the prospect of space exploration. My family has always supported me. I think probably because they are artists they put a very high value on the power of imagination. They always encouraged us to just dream anything."

Because the military was not an option for women in the early 70's, Betsy took the civilian route and enrolled at Parks College of Aeronautics in the fall of 1970. She advanced her ratings through their Flight Technology A.S. program and became the first female to receive a degree from the school when she graduated in 1972. During her years at Parks, Betsy flew on the college flying team and won the "Top Female Pilot" award at the National Intercollegiate Flying Association's Air Meet in 1971. She also won a Beech Scholarship from the Beech Aircraft Corporation that same year.

She landed her first aviation job as a co-pilot on a twin-engine Aerostar flying night freight out of Cleveland, Ohio. "It's usually pretty tough getting that first twin-engine flying job and I was pleasantly surprised that they seemed to have no qualms about hiring an inexperienced women co-pilot for their expanding night freight operation. I learned later that they figured that since I weighed less than the average man, they'd actually be able to carry more freight on the airplane! Whatever the reason, I was logging 8-10 hours a night, five nights a week at $100 per week. I was glad to be building the flight time so quickly but the 12 hour work nights were grueling."

Progressing from the realm of night freight and on to the world of corporate flying was a tough transition and Betsy experienced some disappointments. "Most companies had never hired a female pilot before and once I was even turned down for a corporate job because, as the chief pilot said, "We don't think the other pilots' wives would like it."

When an opportunity to teach at Parks college came available in 1974, Betsy returned to her alma mater as a flight and ground school instructor. A year later, the promise of a left-seat charter position on an Aerostar brought her back to the Niagara Falls airport, the same field where she had earned her private license. The aircraft was sold shortly after she took the position but she stayed on as an instructor.

While reactions were mixed at the sight of a young female pilot, Betsy's congenial personality and professional manner managed to convert skeptics and chronic doubters. "I remember one afternoon a heavyset, middle-aged man walked in and wanted to take an airplane ride. When he learned that the only instructor available was a woman, his face fell and he looked as though he wanted to leave. After many assurances from me and the lineman,he reluctantly agreed to go, shaking his head as he walked slowly out to the airplane. He was quite tense at first, but became more and more relaxed as the flight progressed. By the time we landed and taxied in he was enjoying himself immensely. He left with a broad smile, shook my hand and returned a week later with his young daughter and specifically asked for me."

In 1976, a local charter company started a commuter operation in Erie, PA and needed a co-pilot. Betsy wanted to build more flight time so she convinced them to hire a woman pilot. She got the job and moved to Erie the next day. After a year with the company she realized her career was not progressing as she had hoped and decided to reconsider airline flying.

"In 1977 I enrolled at Purdue University in the Professional Pilot B.S. program. To supplement my student loans, I worked at the local FBO as a flight instructor and charter pilot and occasionally towed banners over the football stadium. Part of Purdue's program included obtaining a flight engineer rating in a Boeing 707. They used an old 707 simulator for the training and the instructors were all retired airline captains. Just studying the old classic airplane and talking to these fellows who had flown the airplane when they were new, made me realize how they felt about their careers as airline pilots and their dedication to the airline tradition. In the course of studying the airplane I developed a special affection for the large transport aircraft and wanted more than ever to be part of a professional flight crew."

After graduation in 1979, Betsy began sending out airline applications. She started with the "A's" and Altair Airlines in Philadelphia, PA was the first to respond. "They were so excited to have a woman pilot within their ranks. The chief flight attendant came a gave me a big hug after my pilot interview and it was really nice to finally be completely accepted." Betsy was their first female pilot. She flew as a co-pilot on the Beech 99 and on the Nord 262 and upgraded to Captain on the Beech in August of 1980. Altair was the first U.S. airline to fly the Fokker F-28 and when the company qualified their pilots to fly the new plane,Betsy became the first female in the United States to earn an F-28 type rating.

Altair ceased operation in November of 1982 but People Express hired her as a B-737 First officer in March of 1983. People Express maintained an innovative labor concept that rotated employees through various administrative positions to enhance their knowledge of the operation. Besides flying, Betsy fulfilled her staff duties by working as a dispatcher four days out of every month. She upgraded to co-pilot on the B-747 in April of 1984 and scored another "first" when she became the first woman to fly a B-747 trans-Atlantic scheduled passenger flight. She upgraded to captain on the B-737 in July 1985 and remained in command until Continental Airlines acquired the company in 1987.

United Parcel Service began its own airline division in 1988 and Betsy was offered a B-757 captain position. "My first night cargo job was such a grueling experience I said I'd never want to fly freight again, but compared to flying passengers in and out of Northeast with People Express, (the most congested aircraft corridor in the country) it seemed rather appealing. Flying at UPS is rather low-key and laid back and I kind of like that now. Cargo planes usually park on the opposite side of most airports so there are no crowded terminals to deal with and no security screening hassles. We just come and go as we please and flying without the traffic congestion I had become accustomed to is a pleasant change."

Breaking new ground and being "first" in so many different aviation areas distinguishes Betsy as a real airline pilot pioneer. The positive image she has maintained throughout her career has been a credit to women in the airline industry and her lighter-side outlook has helped win the war of acceptance. "One night while flying the B-757 with another female first officer, Sue Strausser, I answered one of our radio calls from air traffic control while Sue was momentarily off the frequency. The controller realized he was hearing a different female voice and asked in mock dismay, 'Is that an *unmanned* aircraft?' We laughed and replied that indeed it was."

Besides regular line flying, Betsy became a check airman for UPS in 1990 and gives IOE instruction (Initial Operating Experience) to crewmembers who are new to the aircraft. In her free time, she combines her interest in space and the planet with her love of art and has joined a unique organization called the International Association for the Astronomical Arts (IAAA) This group of artists paints "spacescapes" and travel to different parts of the globe where topography is similar to that found on other planets to practice their craft. "I'm particularly intrigued by concepts such as space-time and the exploration of our relationship with the cosmos through art and myth as well as science." Though not yet a professional in this field, it is only a matter of time before she masters it with the same flare she has achieved in flying.

Chapter Eleven

Air Races

Now that I look back I realize that of all the things the Powder Puff Derby meant to me, everything else was so much more meaningful than winning. Fran Bera

Air racing has incited the adventurous flyer since the First Women's Air Derby in 1929. Famous and exciting races were born in that era and many women won their fame in the spirit of aerial competition. The competition satisfied their ambitions and confirmed their pilot skills as they tested their ability against men, against women and ultimately against themselves.

The Bendix and Cleveland Air Races were dominating events in the 30's and 40's but faded as manufacturers transitioned aircraft development to combat trainers and pilots became involved in war-time activities. Later races including The Angel Derby and the Palms to Pines would have special significance but the oldest, longest and largest air race for women in the world was the All-Women's Transcontinental Air Race, affectionately dubbed the Powder Puff Derby. The evolution of this great race is as much a testament to the women's movement in our country as it is to their undaunted flying ability. Its thirty year endurance did as much to promote the general aviation industry as it did to stimulate its growth.

The Powder Puff Derby was born in 1947 when a Florida chapter of the Ninety-Nines invited their west coast affiliates to fly to Tampa for their air show. Comprised of talented post war WASPs, the Florida Chapter had created an All-Women's Air Show and hoped that a transcontinental air race of women flyers to their event would draw real interest and publicity.

The Los Angeles chapter of the Ninety-Nines rose to the challenge when the invitation was read at their meeting and word quickly spread. Named the

Fig. 11-1 Mardo Crane

Amelia Earhart Memorial Race, only a handful of entrants could commit on such short notice, but, nevertheless, the race was on.

The Derby's first chair, Mardo Crane, recalls: "Weather fouled up the original starting point near San Diego so Palm Springs was selected as the alternate. The first pilot, Carolyn West and her Ercoupe, were flagged off but none of the others ever reached the starting point. She flew cross-country with co-pilot Bea Medes, not realizing they were the only racers until they landed in Tampa 21 hours and 45 minutes later." Blanche Noyes presented trophies to the uncontested winners and a tradition had begun. See Fig. 11-1.

The following year the Florida Ninety-Nines chapter scheduled another All-Women Air Show, this time in Miami, and the efforts to establish an All-Women's Air Race to finish at their air show were once again underway. Mardo organized an actual committee to conduct the event and attracted the publicity and attention the race deserved. Fundraising events subsidized the race and when Jackie Cochran donated $1,500 in prize money, the name was changed to the Jacqueline Cochran All-Women's Transcontinental Air Race.

Each year the route was laid out between two cities that had won bids for the start and terminus of the race, in cooperation with the local Ninety-Nines chapters. Designated stops along the route permitted racers to refuel, wait out weather or remain overnight, as only VFR conditions and sunrise to sunset flying were permitted. Eligible aircraft ranged from single to multi-engine airplanes with varying horsepower limits. Eventually officials imposed a ten year restriction on aircraft to weed out the weary WW II warbirds, as their multiple modifications made fair handicapping impossible.

Timing techniques evolved from the honor system to official cards endorsed by tower operators or airport managers at the end of each flying day. Controversy and criticism regarding inaccurate methods brought about the time clock. This quieted insinuations of cheating, but swelling entries made manual calculations time consuming. Data-processing firms eventually stepped in as technology advanced, enabling race officials to provide daily results to the eager press.

At the Ninety-Nines International Convention in 1950, the membership voted to remove all mention of the All-Women's Transcontinental Air Race from its charter to protect the organization in the event of any unfortunate accident. As a legal, non-profit corporation, the AWTAR (All-Women's Transcontinental Air Race) was still endorsed by the Ninety-Nines, but established a separate identity of its own.

In 1951 and 1952, the Korean War dominated our country's attention so the race was run as a training mission. "Operation TAR" (Transcontinental Air Race) was promoted as a refresher course in cross-country flying for women whose services might once again be called upon as they had been during World War II. No longer just for Ninety-Nines, the race was open to all women pilots and was considered a "race" so long as at least one of the contestants made it past the finish line within the required time limit.

The Derby itself provided perhaps the best training ground ever made available to women pilots. While women sharpened their flying skills to better compete, they demonstrated that safety was attainable through sound flying disciplines. In an effort to increase proficiency and maintain the Derby's impeccable safety record, seminars were held prior to each race. Experts led extensive discussions on subjects specifically designed to the current years route including, wake turbulence, mountain flying, density altitude and desert survival. Also discussed were the effects of flying under the influence of tranquilizers, alcohol and cigarettes. Many Derby racers said, "I learned more flying my first Powder Puff Derby than I learned from all my instructors and books preparing me for the ratings that qualified me to fly it." As an added incentive to increase pilot knowledge and skill, scholarships were awarded for further aeronautical advancement to first-time Derby pilots with the highest scores.

The Derby's popularity inspired women to learn to fly just to participate in the famed race as an inflexible rule required all aboard to be certified pilots. For many it was their greatest ambition to race in the Powder Puff and being a contestant was a way of proving, or even bragging, "Hey, I Can Fly!" Women's achievements in the race won national attention as the Derbies were widely publicized events. Hollywood screen stars, celebrities and other notables took part in the hoopla, even U.S. presidents and first ladies were known to send "good luck" and congratulatory wishes. See Fig. 11-2.

Fig. 11-2 Honorary starter Robert Taylor checks the Operation TAR race route with Mardo Crane, in 1952.

As early as 1950, the pilot look-alike trend emerged as pilot and co-pilot teams distinguished themselves as pairs by wearing matching outfits. Gone were the days of flying suits and goggles, as these modern day trend-setters opted for more fashionable styles. Often the outfits sported their sponsors' logos or matched the paint scheme of their airplane but the team spirit and comradeship that developed from this became as much a part of the race as the competition itself. See Fig. 11-3.

As the race grew and advanced so did the cause of aviation itself as constant reevaluation and revisions promoted safety and insured fair competition for all. Race rules paralleled the basic trends in general aviation deployment and the AWTAR was often referred to as the sounding board of civil aviation.

When race officials discovered that competing racers were flying as high as 20,000 feet to take advantage of brisk tailwinds, they established another rule. This one demanded the use of oxygen above 10,000 feet. The FAA supported the sound-mindedness of this directive and adopted the rule as a requirement for all pilots. Rigid aircraft inspections were also performed at the start and terminus of each race, to insure safety conditions. Planes found not in compliance with airworthiness directives prompted the FAA to establish standardized 100 hour aircraft inspections.

The ever increasing numbers of entries required race officials to stiffen pilot eligibility to include only those with commercial, private with instrument, or instructor ratings. Private pilots could remain as co-pilots, allowing the beginner pilots the ability to build time and experience as second in command. As a result, pilots began advanced flight instruction with fervor to qualify for the Derby and entry into the race became a recommendation of competence in itself.

The race was canceled for the first time in its history in 1974 as the nation's fuel crisis won the patriotic cooperation of the AWTAR board. It was a difficult decision to make but not nearly as devastating as the law suit that was filed against the Derby the following year.

With no transcontinental race for men to fly it seemed inevitable that one would eventually challenge the 28 year "women only" classification. In 1975, a male applicant was disqualified because of gender and he filed suit just three days before the start of the race. Time stood still while the judge deliberated the case and there was considerable concern whether the race could be flown with a male entry. With the race just days away and all plans secured, the effect of a cancellation to sponsorship would have been disastrous. To the relief of the race board and pilots alike, the judge ruled in favor of the all-female tradition and the race flew as scheduled.

Applauded for their careful management and sensible rules, the six to nine woman race board was an efficient organization that dutifully completed the myriad of details required to keep the Derby winging across the country each year. This planning was like piecing together an intricate puzzle, and the assembly of these separate pieces required year-round attention. Their dedi-

Fig. 11-3 Winners of the 1973 Powder Puff Derby, (r) pilot Sylvia Payton and co-pilot Pat Fairbanks show their matching trophies and outfits.

cated efforts as well as the untiring energies of thousands of volunteers led the Powder Puff Derby to prominence.

Each year their task grew more difficult. Designing the routes and coordinating with the authorities became more perplexing as airport terminal control areas grew larger. The struggle of finding accommodations for hundreds of racers and the endless search for funds and sponsorship finally drove the AWTAR to announce that the 1976 race would be the last official Powder Puff

Derby. A time honored tradition was coming to an end and shockwaves went through the aviation community.

Anticipating the deluge of entries for the Derby's last hurrah, the board secured accommodations for 150 teams. To their surprise, they received 235 properly postmarked entries on opening day and arrangements had to be adjusted. From Rhode Island to Rhodesia, the women came to race. Sixty teams were made up of those who had never flown the Derby before, and the others, old pros who dreamed of one last chance at winning it. A record 16 teams were mother and daughter duos, proof that early enthusiasm for flying begins at home. As a fitting finale to the 29 year Derby, the race would be flown from "Capitol to Capitol." Sacramento, CA to Wilmington, DE, a total of 2,926 miles, the longest and the largest race. See Fig. 11-4, 11-5, 11-6. 11-7.

When AWTAR officials traveled to the National Air and Space Museum to engrave the final winner's name on the Derby trophy enshrined within their halls, the Museum's deputy director, Michael Collins remarked at the shame of terminating the race at number 29. He suggested one more to make it an even thirty and proposed the race fly the original 1947 course. Through his encouragement, the "The Smithsonian Milestone of Flight" was planned as the 30th Commemorative Powder Puff Derby, to honor the thousands of women who had successfully flown from coast to coast for 29 years.

Race officials went back to the drawing board and completely revised previous race rules to include a broader spectrum of pilots and planes. Speed would not be a factor, eligible aircraft could range from 65 - 600 horsepower and any passenger who had once attained pilot status was allowed to ride.

The Derby's original winner, Carolyn West, was on hand to bid everyone good luck, and Jackie Cochran braved the sweltering 127 degree temperature on the tarmac at Palm Springs to flag off the first plane. Pilot pioneers Viola Gentry and Louise Thaden waited for the flyers at the finish line in Tampa, and Blanche Noyes was in place for the awards presentation as she had been 30 years before.

Enthusiasm at each stop seemed greater than ever for the 331 pilots who made the sentimental journey. Among them were 52 grandmothers, seven FAA flight examiners, and three flying the race for the 24th time. In all there were 55 different makes and models flying to the finish line.

Thirty years of Powder Puff flying proved that there was certainly more flying than puff. There were more than 4,400 contestants who flew 638,536,800 miles, the equivalent of more than 225 times around the world at the equator, WITHOUT A SINGLE FATALITY! This was an amazing feat considering that at times as many as 200 race planes were flying in the same airspace and aiming for the same airport at the same time. Race chairman Kay Brick proclaimed, "This in itself is a tribute to the organization's emphasis on safety, the dedication of the women racers and to the diligent efforts of the FAA controllers who handled the fly-bys and supported our derby events year after year."

Fig. 11-4. Bonnie Belle Whiteleather (Goddard) is seen here sprinting to punch the time clock.

Fig. 11-5. (l) Mother and daughter team Bonnie Seymour and daughter Linda Seymour in 1976.

Fig. 11-6. (r) Mary Ann Noah, pilot and Peggy Wright, co-pilot in their winning entry, "Noah's Lark."

Fig. 11-7. Dottie Bock (l) and Ellie McCullough shown here in 1968. Their plane says it all.

Never would there be a better showcase for general aviation. The women pilots and their Derby had proven that flying was safe, reliable and fun. FAA administrator Langhorne Bond bestowed a baccalaureate on the Derby with these words: "The Powder Puff Derby was a grand event carried off with unusual expertise and professionalism, and most assuredly will rank with those aviation feats of our biplane pioneers."

When the final AWTAR was flown in 1977, another group of dedicated pilots formed a Board of Directors and successfully ushered in the Air Race Classic. Their first race covered 2,606 statute miles and was flown July 16-19, 1977, from Santa Rosa, CA to Toledo, OH.

The Air Race Classic has developed into the current, prime cross-country race. The first fifteen ARCs have averaged 2,489 statute miles over the race course, attracted 71 pilot-entries in 37 aircraft. The participants continue to perfect their flying techniques and enjoy the camaraderie. In 1993, the seventeenth ARC starts from Corpus Christie, TX and terminates at North Kingstown, RI June 22-25.

Mardo Crane

Mardo Crane first flew in a barnstormer's airplane in 1924 at the age of 16 and learned to fly at the suggestion of Amelia Earhart in 1931. She received her bachelors degree from UCLA, her masters degree from Columbia University and completed studies in both Drama and Radio Broadcasting as well. She made a living writing and broadcasting for KROD in El Paso, Texas and taught flying to students on the side.

Mardo entered the Women Auxiliary Ferry Service in 1943 where she served for seventeen months. Besides her ferrying duties, she towed targets, flew as a test pilot and flew strafing, radar and searchlight missions. Among the mighty machines Mardo was qualified to fly were; PT-19, BT-17, UC-78, AT-6-7-13-11, PQ-14, the Douglas Dauntless and the Martin Marauder-B-26. "The biggest thing I flew was the B-26 and after intensive training and towing ground-to-air and air-to-air targets at Las Vegas, I was transferred to March Field in California where there wasn't ONE B-26 on the base. That was the Army!"

After the sad and untimely release of the WASPs in December of 1944, Mardo was deactivated and became an editor of Aviation News. It was Dianna Bixby who was originally asked by the Florida Ninety-Nines to help organize a transcontinental women's air race and she turned the job over to Mardo.

As pioneer race chairman, Mardo set a course that would turn the Powder Puff Derby into the longest running air race for women in the nation. "We called it the All-Women's Transcontinental Air Race but the press would have none of that after Will Rogers had dubbed it the Powder Puff back in 1929." The name stuck as did the traditions that were born of those early races years.

"It was always a struggle to get the prizes and the sponsors and I never understood why the airplane industry didn't back us more." Remarked of one sponsor, "Mardo I don't know what you are trying to prove, women showed a long time ago that they could fly."

"The race grew as time went by until over a hundred planes became the norm. It was a fun race and we did what we set out to do; focus on women's ability to fly, show that flying was a safe means of transportation and that private flying, (now called general aviation) could bring attention and business into a town."

"The Powder Puff Derby was a wonderful experience for all who participated in ANY way at all. To have had a part in the race is to have been included in an exciting historical event." Mardo was the Derby's race chairperson from the start of the race in 1947 until she turned the reigns over to her successor in 1952. She remained active, supporting the AWTAR organization and even flew in the famed race in 1969, 72 and 77.

Mardo was the first and only woman ever awarded the "Barnstorm Pilot of the Year" trophy by the Flight Museum of Lancaster, California and is notably listed in Who's Who of American Women, Who's Who in Los Angeles County and Who's Who in Aviation and Space.

As an active participant in two of the most historic periods for women pilots, Mardo brought her experiences to life by penning two novels. *Flydown of the WASP* and *Ladies! Rev Up Your Engines!* Today Mardo makes her home in California were she is a graphoanalyst, (handwriting expert) an interesting sideline she has been active in for more than 10 years.

Edna Gardner Whyte

Edna Gardner Whyte was 18 years old when the 19th Amendment passed which gave women the right to vote. "Why should women have to beg? said Edna, "Why can't they do anything that a man can do?" And during her 57 year career in aviation, Edna did almost everything a man could do.

Injustice flamed her spirit and personal hardship toughened her. At the age of eight she lost her father. A year later her mother was hospitalized with tuberculosis and was forced to give her three children to relatives and friends. For Edna's ten most important and formative years, she was moved from house to house.

"I hated being poor," she said, "and I hated feeling unwanted."

Edna studied nursing and became a registered nurse. She went for her first plane ride in 1927, while working at the Virginia Mason Hospital in Seattle, near Renton Field. It was right after her first plane ride that Edna ran into her first problem in aviation. Everywhere she went the male instructors discouraged her: They told her, "Give it up. You're going to kill yourself." As a result she let her student certificate lapse. Finally she found a male instructor willing to teach her to fly. She fell in love with flying and it became the center of her life. Nursing was able to finance her new passion.

Edna had read that the Stinson sisters had taught men to fly combat for World War I, and she figured, "If those two slips of girls and others that I'd read

Fig. 11-8 Edna Gardner Whyte

about - Ruth Nichols, Phoebe Fairgrave Omlie, Anne Lindbergh, Amelia Earhart and others could do it,why couldn't I?" Edna had dared to dream!

Edna joined the U.S. Naval Nurse Corps to help survive the Great Depression. She soloed in a Swallow T.P. in January 1931. In May, her pilot examiner said, "Lady, I haven't given a certificate to a woman before and I don't see why I should start now." She cried and he relented.

Edna got her private license in 1931, and joined the Ninety Nines. Edna was later to become president of the organization 1955-1957. She also became a charter member of the Whirly-Girls (#10). See Fig. 11-8.

Four months after obtaining her private license, Edna decided to join the air racing circuit. She showed up at the Cleveland Air Race in her Travel Air. Since her plane was not a racer, she teamed up with parachutist "Cowboy" LaPierre. She would drop LaPierre from the lower wing and had plans to share the week's income that his stunt would bring.

"At the end of the race week, I was thrilled," said Edna. LaPierre had landed in the painted circle every time he jumped. "I figured he had earned a lot of money for us. After the last jump he rushed to the air race officials while I parked the airplane. By the time I worked my way back from the field he was long gone. He absconded with the money and left me flat broke! I had to borrow gas money to get my Travel Air back home. I watched for that man for the rest of my life."

Whyte was a product of the "Golden Age of Aviation." She rubbed shoulders with the famous and not so famous men and women pilots and grew from the experiences. She returned to the air races for many years and was able to compete in races through a combination of borrowing, bartering and cunning. She began hauling passengers, stunting airplanes and soon became a premier air racing pilot. She once borrowed an OX-5-powered Waco and won the 1933 Kate Smith Trophy Race. A year later she flew a J6-5 powered Aristocrat to win the Annette Gipson Trophy Race.

When the Navy transferred her to Washington, Edna began teaching students, ferrying passengers and continued racing planes. She added to her repertoire precision landings, bomb-dropping balloon bursting, cross-country and closed pylon racing.

Her disappointments and heartaches sometimes overshadowed her achievements. In 1934, flying her Aristocrat, she beat eight male pilots in a race at Curtiss Field in Baltimore, Maryland. The officials grudgingly gave her the trophy but the following year when she showed up to race again she found the race directors had posted a notice, "FOR MEN ONLY." She won races at the All American Maneuvers in Miami, Florida in 1935, 1937, and 1938-1941. Yet she was greeted not with cheers and well wishes but taunts. "Why don't you stay at home with diapers and dishes," the insecure men would say.

Her first marriage, to Ray Kidd, resulted in her leaving the Navy and a combined effort to operate a flight school. Edna obtained her seaplane rating and taught many men to fly on land, and water at the New Orleans Air College. She and her husband also operated a Civilian Pilot Training Program prior to the war. The marriage ended in a bitter divorce, but that did not stop Edna from her dreams. She went on to get her instrument rating and to teach instrument flying to the male military pilots at the Spartan School in Tulsa, Oklahoma. When the World War II pilot requirements tapered off, Edna joined the Army as a nurse and spent the rest of the war in the Philippines.

In 1946, Edna married George Whyte, a pilot/mechanic, and for the next two decades Edna Gardner Whyte won races in a highly modified SNJ. She also flew in competition at hundreds of events all over the country. She went on to become one of the premier stock plane racers in the Women's Pylon Racing Association. She also spent more than 13 years as a saleswoman for Channel-Chrome, a company that specialized in the chroming of engine cylinders. Edna flew thousands of air miles covering almost every country in the Western Hemisphere. The chroming of her engine's cylinders increased the speed of her airplanes and the company sponsored her flights, making her entry into dozens of racing events possible.

Edna Gardner Whyte's living room bulged with trophies. By the time Edna was 80 she had won more than 125 air races and stunting trophies. She received the Charles Lindbergh Lifetime Achievement Award in 1986, and the Godfrey Cabot Award of the Aero Club of New England in 1988. She was the first woman elected as a Honorary member of the Order of the Daedalians, and she became the 10th woman in the free world to earn her helicopter rating, in 1951.

Her 20 year marriage to George Whyte ended in divorce but that heartbreak did not stop her either. By 1971, at an energetic 69 years old, Edna had built an airport and had workmen carve a runway and build a hangar in which she lived until she was 80. She continued to pump gas, teach students to fly, drop skydivers, teach aerobatics and fly in air races.

Amelia Earhart once told Edna, "Women faced stone walls and heartaches."

"Although she was referring to women in aviation, I faced those obstacles from my earliest years. Instead of dissuading me from pursuing goals as a pilot, obstacles seemed to fuel my competitive spirit and Amelia's words and actions inspired me to make aviation my career."

Edna was a member of the Silver Wings, the United Flying Octogenarians and the former president of the Ninety-Nines. Edna took her last flight on February 15, 1992, and she was to have been honored in 1992 as one of the Eagles at the Gathering of the Eagles, at Maxwell Air Force Base. She was just nine months short of her 90th birthday.

She reluctantly relinquished her role as pilot-in-command in 1986, when she lost her medical certificate. Throughout her life Edna Gardner Whyte demonstrated the ability to rebound from adversity, and to create something worthwhile despite confrontation. She continued Rising Above It.

Fran Bera

Fran Bera started flying in high school at Grand Rapids, Michigan in 1940. After qualifying for her commercial license and flight instructor rating, she built her flight time ferrying surplus trainers around the country after WWII. She progressed from flight instructor of airplanes and helicopters to designated pilot examiner for the FAA and licensed more than 3,000 pilots in her 25 year career.

As an airline transport rated pilot, Fran is also qualified to fly seaplanes and hot air balloons and has logged more than 25,000 hours of flight time. She has been a chief pilot, a charter pilot and a flight operations manager and has operated her own flight school and aircraft sales business. Fran also flew as an experimental test pilot for Lift Systems, Inc., a company who developed new designs in rotorcraft. Fran also spent time as a free fall parachutist.

When Fran moved to California in 1950, her passion became air racing, an interest that dominated her life for 20 years. She competed in the International Women's Air Race, The Reno National Air Races, the Angel Derby and in The Great Race from London, England to Victoria, B.C., but won her fame as the Powder Puff Derby's winningest pilot, streaking across their finish line an unprecedented seven times. See Fig. 11-9.

Fran raced the Powder Puff for 19 consecutive years from 1951 to 1969 and then again in 1976 and 1977. She placed in the top 10 seventeen times during her 21 races, (5 times in second place) but nothing but winning would satisfy her. "After a couple of wins people began to expect it. That put a great deal of

Fig. 11-9 Fran Bera

pressure on me. After coming in second by only 34 seconds they said 'What happened to you?' So nothing but first place counted. I was very competitive and winning seemed to be everything." See Fig. 11-10.

While everyone's combination for winning included having a good pilot, a sharp navigator, a fast plane and strong tailwinds, all eyes were on Fran as the woman to beat and special significance came to those that could beat her. Among the Derby's jingles was a ditty arranged by 8-time Powder Puff racer, Hazel McKendrick Jones, who put contestants-thoughts to music in this excerpt sung to the tune of "Oh My Darling Clementine,"

Mr. Weatherman, Oh Mr. Weatherman
Will you help us if you can
We need weather, real good weather
'Cause we'll all be following' Fran...

"The AWTAR meant other things to me, such as making friendships with other women pilots that have lasted a lifetime. These were women from all parts of our country and the world. They understood the pure joy of flight and the beauty of seeing our country unfold from coast to coast under our wings, while we tested our skills to the utmost. It meant the excitement of competing. We

Fig. 11-10 Some of Fran Bera's many trophies.

took off into the rising sun on a still morning, and at days-end we shared the fun and companionship while telling our stories of the frustrating minutes we had lost. We shared the humorous things, the bad breaks, the good tailwinds and making it just past the finish line ahead of thunderstorms that would close the field. It was wonderful to communicate all of this to other women and know they understood. Now that I look back I realize that of all the things the Powder Puff Derby meant to me, everything else was so much more meaningful than winning."

From race pilot to record holder, Fran also holds the world altitude record for Class C-1-d which she set in 1966 in Long Beach, California. She managed a phenomenal 40,194 feet in a Piper turbo-charged Aztec, a record which still stands today.

The FAA organized a special Women's Advisory Committee that Fran served on for three years, as well as on Governor Reagan's Aviation Education Task Force in California. On March 4, 1975 Fran's name was written into the

Congressional Record in "A Salute To Women In Aerospace" by the Honorable Don H. Clausen. It was a special recognition for her dedicated efforts as an aviation pioneer and the leading role she has played in advancing aviation, aeronautics and aerospace programs and sciences.

In 1978, Fran's name was engraved in granite and added alongside other famed pilots in Memory Lane, at the International Forest of Friendship in Atchinson, Kansas. In 1980, Fran was named "Woman Pilot of the Year" by the Silver Wings Fraternity.

Fran is an active member of the Ninety-Nines, the Whirly-Girls and other aviation organizations. She retired in 1985 after a successful career as a sales representative for both Beech and Piper Aircraft.

Fran was seriously injured as a passenger in a Lockheed Electra that crashed at Oshkosh in 1990. She miraculously survived and is on the road to recovery with the same determined spirit that propelled her across Derby finish lines, and into the history books as a Powder Puff legend.

Susan Dusenbury

The first Women's Air Derby was organized as the opening attraction to the 1929 National Air Races in Cleveland, Ohio. Never before had women pilots been assembled for the sole purpose of competition. To the women it was more than just a race to the finish line, it was an opportunity to establish women's importance in the development of aviation and to prove that they were capable and competent flyers. Upon the successful completion of the grueling event even skeptics were forced to accept women as serious pilots. Generations would long remember the glorious race and the spirit and sense of adventure these pioneering women possessed. In August of 1989 a sixtieth anniversary commemorative tour was scheduled to honor the derby's first winner, Louise Thaden and the other 19 contestants. A core of volunteers was assembled and hours of research went into mirroring the accounts of the original race. W.C. "Dub" Yarbrough, President of the Staggerwing Foundation and close friend of Louise Thaden, assisted in the efforts. His task was to choose a pilot to fly the winning 1929 Travel Air D4000 "Speedwing" along the original race route and he placed the honor in Susan Dusenbury's capable hands. See Fig. 11-11.

Susan started flying when she was sixteen in a friend's Piper Colt and soloed after just four hours of instruction. She was a natural pilot and decided early that she wanted to make aviation her profession. She completed her ratings and flight instructed between school semesters to build time.

Susan became the first woman in the United States to earn an inspection authorization rating after receiving her degree in aero-maintenance from Florence Darlington Tech in South Carolina. She also received her bachelor's degree in business administration from Florence, South Carolina's Francis Marion College.

She remained at Tech for two years as an associate instructor of airframe and mechanics before opting to fly for Air Carolina, a commuter airline in the

Fig. 11-11 Susan Dusenbury

Southeast. She went on to fly Navajos and Metroliners for Air Virginia for five years and then King Air 200's for the State of Virginia. There she flew for the Department of Industry Development and occasionally for the governor. In 1983, the state honored Susan with special recognition for outstanding achievement and dedication to Aviation.

Airborne Express hired Susan in 1985 and today she flies as one of their DC-9 captains. Next to flying, restoring and repairing antique and classic airplanes is Susan's passion. Her 50' x 70' hangar houses some of her restoration projects which include a 1930 Inland Sportster, a 1930 Fairchild KR-21 and a Culver LCA "Cadet" which won her Oshkosh's Restored Antique Airplane Award in 1988.

As a dedicated member of sport aviation, Susan co-founded the EAA Chapter #629 in Florence, South Carolina, and served two terms as an EAA president. One term for Chapter #8 in Greensboro, North Carolina and another for Chapter #646 in Roanoke, Virginia. This devotion and service earned her EAA's Major Achievement Award in 1983. She is a member of the Ninety-Nines, the Staggerwing Museum Foundation, The First Flight Society of North Carolina and Virginia's Aeronautical Historical Society. She was recently elected a Class III Director of the EAA.

Such credentials made Susan an obvious choice as the commemorative flight's pilot. "It was an honor to be selected to fly the route. Louise Thaden had always been the woman I admired most and I felt this was an opportunity to give her the recognition she deserved." Susan helped turn the dream into reality by using her licenses in airframes, powerplants and inspection to complete the necessary paperwork needed to get the derby's blue and gold biplane airworthy.

While Susan worked on the airplane, the tour committee used a 1928 Rand McNally road map to locate original landing sites and selected alternate stops where airports no longer existed. Newspaper accounts from the original cities and interviews with surviving witnesses helped to duplicate the events of 1929. Louise Thaden's son, Bill, had the most difficult assignment. He coordinated the details with ATC so the vintage Travel Air with no radio or navigational equipment could fly through modern controlled airspace.

Finally as the August 18th departure date approached, details were finalized, plans were complete and both the pilot and the plane were ready. On July 26th, Susan took off from her hometown in North Carolina to reposition the airplane for the Santa Monica start. She made a pit stop enroute to California to display the famous aircraft to all those in attendance at Oshkosh '89.

Susan dressed in authentic flying gear and convincingly fit the part of a 1929 racer. "I wore a white camp shirt and riding britches or jodhpurs. I was even able to locate a pair of riding boots that had been handmade in London in 1923. They proved to be every bit as uncomfortable as they were authentic." Flying goggles, helmet and Louise Thaden's silk scarf completed the ensemble. Sun screen was the only item in Susan's gear that would have been nonexistent in 1929.

Original race pilots Mary Haizlip and Bobbi Trout were among the crowd of well wishers as was Mrs. Cliff Henderson, the widow of the famed National Air Race Promoter. Pilot and film star Cliff Robertson served as Master of Ceremonies and fired the shot that signaled Susan's departure.

Activities were planned at each stop thanks to the collaborating efforts of local Ninety-Nine chapters, airport directors and fixed base operators. Eighteen cities in all pulled out the stops to provide a worthy welcome with waiting crowds, waving banners, keys to cities and hangar barbecues. "There were so many folks that turned out at the airports for our arrival. In Arizona a man who was nine years old back in 1929 came out to meet the same flight sixty years later. He remembered how excited everyone was to see those women flyers and the way they were dressed."

While the tour committee was successful in their attempt at making time stand still, aviation's evolution would prove otherwise. The original women racers charted their courses with road maps and flew from point A to point B as the crow flies. Susan was forced to fly a more precarious route using sectional charts to maintain separation from today's antennas, towers and restricted airspace.

There were no TCA's to contend with back in 1929, and without a radio for ATC communications, Susan had to rely on her chase pilot, Brad Thomas, to pass her in traffic areas and receive instructions she could follow. Traffic and congestion found at present airports serve as a contrast to yesterday's open fields with single runways. "Back in those days the women racers could easily take off towards their destination to save valuable race time. I had no desire to break Louise's winning record of 19-plus hours. In fact with present constraints I actually added about four hours to the route."

At the successful completion of the commemorative tour, Susan put her accomplishment into perspective. "The tour symbolized the dramatic achievement that women have made in aviation. Louise Thaden was more than just the winner of that first race, she represented the spirit of all women aviators,their ability and their courage."

Marion Jayne & Nancy Palozola

As the winner of more cross country air races than any other pilot, male or female, in the world, aerial ace Marion Jayne Palozola and her daughter Nancy competed in the first ever around-the-world air race in 1992. See Fig. 11-12

The grueling 15,000 mile race circumnavigated the globe, flying from Geneva, Switzerland to Cannes, France by way of airports in Finland, Russia, the United States, Canada and Greenland. Enduring more than 175 hours in five weeks to complete the round trip course, Marion reveals that it was by far the toughest race she had ever flown. "Mentally and physically it was a draining race because we flew 13-hour, non-stop legs, mostly at night, through all sorts of weather conditions ranging from severe icing to thunderstorms."

Top pilots from all over the world came to compete. Of the 28 teams that entered the race, all were men except for one other female Pilot-In-Command whose mechanic was aboard, and the Palozola's all-female team. "We competed against pilots from France, Norway, Switzerland, the United States, Guatemala, and Hong Kong and the camaraderie was outstanding. Men and women worked together and helped one another by giving position reports throughout the race. In fact from Helsinki to Moscow there were thunderstorms along our route and we were making position reports about every 50 miles. The air traffic controllers were a little overwhelmed with all of us and at one point announced over the radio: "All race aircraft, stop where you are!" I imagine there was quite a bit of chuckling in the cockpits when that message was received."

Treated in each country like celebrities, the race pilots were wined and dined as guests at the French Embassy, they were granted permission to fly over the

Fig. 11-12 Nancy (l) & Marion Jayne Palozola

newly opened Russian airspace and landed at a secret airport outside Moscow. To the press, the sight of women flying around the world gave pause and one reporter questioned Nancy asking: "Why aren't you home with your babies?"

Nancy, as well as her sister Pat, learned to fly under their mother's instruction and frequently co-piloted in Marion's race adventures with winning success. Marion and Nancy reigned as the world's undefeated mother/daughter air race duo and together their mission was to show the world that air racing was something that women too could do. Marion boasts that her daughters are two of the best navigators there are in aviation. "They do all the radio and navigation work so that all I have to do is concentrate on flying the airplane. Races are often won and lost in a matter of seconds. I lost one race in Nicaragua by 22 seconds and can't help but think of all the places I could have saved the time."

The Palozola's winning combination of teamwork and flying skill won them four of the round-the-world legs. In the final stretch they led the event by two minutes but were forced to make a final stop in Ireland, for fuel. Unable to make up the time, the previously undefeated mother/daughter team landed second. For them there was no agony of defeat, just the thrill of competing. The Palozolas were happy with their second place standing. They are now gearing up for a challenge of greater grandeur in a 25,000 mile around-the world race scheduled for 1994.

Air racing is not the only venture for this winning team has succeeded in. They are also the co-founders of Tailwinds, the aviation catalog of the Skies. Plotted on the same kitchen table as their air race routes, they sought to market aviation products to others who share their "spirit of flight." As their popular catalog company celebrates its sixth anniversary, Marion and Nancy have crossed yet another finish line with flying colors.

Chapter Twelve

The Civil Air Patrol

Prior to World War II, the aviation industry did little to encourage women. At the outset of the war, of the 100,000 licensed pilots at the time of Pearl Harbor, about 3.5 percent were women. many of these women wasted no time offering their services to the Civil Air Patrol, when it was formed in December 1941.

When the Civil Air Patrol couriers spread their wings to carry urgent shipments to defense plants and Army depots, women pilots of the Patrol shared the flying responsibilities alongside the men. The women also flew on missing plane search missions, mercy missions and forest patrols. Although they were not permitted to fly coastal or Mexican border patrols, each CAP unit boasted from three to a dozen women who helped keep the CAP flying.

At the outset of the war, women with aviation backgrounds were enlisted by the CAP along with male pilots and technicians. There was little discrimination because of sex. Individual ability, experience, and their past record counted. The first national commander of the Patrol, Major General John F. Curry, encouraged women when he announced: "There must be no doubt in the minds of our gallant women fliers that they are needed and, in my opinion, indispensable to the full success of the CAP. A great part of the progress made in organizing civilian aviation under the Civil Air Patrol has been due to the volunteer help given by women fliers - members of the Ninety-Nines, and the Women Fliers of America."

The women who joined the CAP represented every occupation in America. Schoolteachers, doctors, artists, writers and college students joined the ranks with secretaries, telephone operators and housewives.

Some of Hollywood's shining stars joined; among them Mary Astor, who went on to become a plotting-board operator at a coastal patrol base; Carol Landis, a volunteer nurse attached to the sheriff's squadron in Los Angeles, and Joan Fontaine, who flew her own plane.

When Cornelia Fort returned to the United States after Pearl Harbor she joined the Texas Wing staff of the CAP and encouraged enrollment of women with her speeches and flying demonstrations. Fort was the second person to volunteer for duty with the Women's Auxiliary Ferrying Service.

At some locations, women formed their own flights and squadrons; in most instances, however, they belonged to the same units as the men. In either case, the women senior and cadet members tackled the same training courses as the so-called stronger sex and received appointments as officers and noncoms.

The attitude of most of the officers regarding women in the ranks of the CAP was voiced by the commander of the Los Angeles group in the summer of 1942, when training was at its height. Referring to the newly formed women's flight, he said: "They'll know how to fix their own planes by the time they finish their mechanical training. And they'll master first aid, drill, meteorology, and navigation. They'll spend two hours in the drill hall and classroom nearly every night in the week. They'll learn all there is to know about mountain terrain and fire-spotting. Then, in August, when the flight unit is complete, they'll be ready to take their places alongside the men in the Patrol."

Of all the CAP women who ever climbed into a cockpit, one of the most beloved and colorful was Second Lieutenant Maude Rufus, a public relations officer of the Ann Arbor, Michigan, squadron. She was known as the "Flying Grandmother" because she had soloed at the age of 65, long after the birth of her first grandson. Rufus was a member of the Ninety-Nines and the author of a book called, *Flying Grandmother, or Going Like Sixty.* Mrs. Rufus accumulated more than 1,000 hours and was a veteran cross-country flier.

What was proper dress for women pilots had been an unresolved issue since Harriet Quimby first raised it back in 1911. The CAP, like many organizations women were joining, was not without controversy around what was appropriate dress for its women pilots. The issue was over slacks or culottes and what was a practical and attractive flying uniform for the women.

One women wrote with conviction: "Our squadron believes that slacks are more practical than culottes and less encumbering to girls climbing in and out of planes and wearing parachutes. We have never seen any girl wear culottes while flying. Slacks and jodhpurs have always been the dress."

Another woman countered: "Most girls and women look horrible in slacks, no matter how well-tailored - unless they are John Powers models. If the women in the CAP are to wear uniforms, why don't you include optional culottes which are just as practical for flying and a lot more sightly on the ground?"

The National Supply Officer, Harry Playford, resolved the issue with the wisdom of King Solomon. He proclaimed that culottes were to be the official uniform for street wear and that a squadron could select its own flying togs, subject to approval from the national headquarters.

When the WASP were disbanded in December 1944, they were welcomed with open arms into the CAP. A CAP official said, "All over the country, the WASPs are coming back into local aviation. These girls representing thousands of dollars' worth of expert technical training, are rendering extremely valuable service to the CAP. It's a real homecoming for many of them."

The wartime experiences of the women in the CAP deserves to be told. Lieutenant Clara Livingston of a Cleveland CAP unit distinguished herself on a search mission for a crippled lake barge in Lake Erie. With her wings loaded with ice, and flying through a blizzard that cut visibility to almost zero, she located the barge for the Coast Guard. After risking her life to locate the barge, and notifying the Coast Guard she stayed on station until she was nearly out of fuel. In a sad note of irony, the Coast Guard was unable to reach the barge in time and it went down with all hands.

On one forest patrol mission, two women flying as observers spotted arsonists and notified the authorities. They continued to fly the mission and played a part in the capture.

Women mostly flew as couriers, some in the mountainous west, but many carried cargo for war plants in the Great Lakes region and throughout the industrial east.

Not all the CAP stories ended happily. Some women pilots gave their lives for the Patrol and their country. They always paid close attention to fundamentals and to the fine points of flying. While the pilots seldom took unnecessary chances and guarded against indifferent maintenance or any other factors that could lead to trouble, several women were killed when overtaken by bad weather.

Willa C. Brown Chappell

A decade after Bessie Coleman, the first licensed black pilot died in a tragic accident, another black woman would gain a notable a position in aviation. That woman was Willa Brown Chappell. She was not the flamboyant barnstormer that Bessie Coleman was but her impact on aviation was significant.

Willa Chappell earned a Bachelor's degree from Indiana State Teacher's College, and migrated to Chicago in 1932. It was there that she heard of the first black woman to fly. When she learned that Bessie Coleman had died, she searched for someone who would teach her to fly. She felt strongly that another woman carry on Bessie's campaign to encourage blacks into the field of aviation.

Chappell went on to earn a master's degree from Northwestern University in business education but at the same time began studying aviation. She studied airplane mechanics in the Aeronautical University in Chicago's Loop, between 1934 and 1935. Chappell went on to earn a Master Mechanic Certificate. By this time there were schools that would teach a woman to fly and she took her flight lessons out of the Harlem Airport in Oaklawn, Illinois. She earned a commercial

Fig. 12-1 Willa Brown Chappell

pilot's rating, and a CAA ground school instructor's rating. Her most notable achievement was in establishing two aviation programs including one taught to college students interested in flying. She taught ground school at night and then took her students to Harlem Airport where she gave them flight lessons.

Chappell, like many Americans, saw the war clouds coming and in 1940 persuaded the Civil Aeronautics Authority to accept that her students proved that blacks could become qualified pilots and be accepted into the Army Air Corps.

Once she overcame that seemingly insurmountable hurtle, Willa Chappell then campaigned successfully for the Army Air Corps to train blacks as pilots.This led to the legendary program at Tuskegee Institute.

Willa Chappell married Lt. Cornelius Coffee, also a pilot and they opened the Coffee School of Aeronautics in 1942. Bessie Coleman's dream had finally been born. The Coffee's School was the first formal flying school owned and operated by blacks and approved by the U.S. government. This school was responsible for the initial training of the men who became pilots in the 99th Pursuit Squadron the highly decorated all-black fighter squadron of World War II. See Fig. 12-1

In 1942 Willa Brown Chappell joined the Civil Air Patrol, and became the first black woman to become a member of that organization. Brown, went on to organize Squadron 613, of the Civil Air Patrol, and was the first black to hold the rank of officer in that organization.

After the war, Chappell was an unsuccessful Republican candidate for Congress several times through the late 1940s. She attempted to establish a Chicago-area airport owned and operated by blacks and made its creation part of her political platform.

Willa Brown Chappell taught aeronautics in Westinghouse High School, in Chicago until she retired in the mid 1970s. She took her last flight at the age of 86, on June 20, 1992.

Colonel Nanette Spears - Wing Commander

In 1956, Colonel Nanette M. Spears became a CAP Wing Commander. This no nonsense leader of young CAP cadets became a legend during her 13 years as Commander of the New Jersey Wing of the CAP. At the time, she was one of only three woman wing commanders. She developed a loyal following of cadets and officers and became known affectionately as, "The Dragon Lady."

Nanette was a striking brunette who smiled frequently and was constantly traveling to hundreds of CAP and other aviation functions throughout the state of New Jersey. She had one mission, to preach the importance of aviation education to high school and college students.

Fig. 12-2 Col. Nanette Spears

Nanette Spears did not have aviation as an early goal. At 17, she was a concert pianist, and a graduate of the Cincinnati Conservatory of Music. In her early years she had her own radio program.

In the mid 1930s she married Albert Spears and became an accomplished horse woman who owned several vigorous hunting horses as well as polo ponies. Her flying years did not begin until 1939.

"I first took up flying as a sport in 1939, while on a visit to California. My husband had not been happy with my previous hobby, which was playing polo and riding horses daily. He did not think it was safe. When I wrote to him about my new hobby, he sent me a 22-page letter trying to dissuade me.

The letter did not deter Nanette and when she returned to her home in New Jersey she continued to take flying lessons. After she'd been flying a year, the airport manager asked her if she would answer letters he was getting from people wanting to learn to fly. She agreed and he set her up in an office with a desk.

Nanette worked at this job as the war clouds in Europe began to descend on the United States. Nanette followed the papers about the proposed civilian flying organization that would perform certain aviation activities in the event of war. On December 1, 1941, just one week before Pearl Harbor, the President of the United States signed the bill authorizing the Civil Air Patrol. Nanette was one of the first to sign up and became its first recruiting officer.

Nanette Spears quickly moved up the ranks and served at all levels working in every capacity in the Patrol, from squadron to wing. Just before her promotion to Wing Commander Colonel Spears earned the Civil Air Patrol's highest decoration, The Distinguished Service Medal. The citation read,

Colonel Spears' activities with this wing show a record of continuous and uninterrupted service since the very inception of the Civil Air Patrol. This officer has continually demonstrated exceptional ability and through her exercise of logical reasoning and sound judgement has become the keystone to the administrative structure of the entire wing.

Colonel Spears is never hesitant to accept duties of great responsibility and, in the assumption of these duties, has repeatedly demonstrated an exceptional degree of individual and moral stamina which has contributed in high degree to the success and steady growth of this wing.

In 1962, cartoonist Milton Caniff, the creator of the comic strip character "Terry and the Pirates" met Spears. One of the leading characters in the comic strip was the "Dragon Lady." Ironically, Caniff had drawn the Dragon Lady in the likeness of Colonel Spears. Caniff presented Nanette Spears with one of her most prized possessions, a painting of the Dragon Lady, with the inscription, "The C.O. is the D.L." See Fig. 12-2

A year after Spears stepped down as Commander of the wind she passed away. Following her death, the New Jersey Wing adopted the name, "The Dragon Wing" in her honor.

Colonel Virginia E. Smith - Commander - Rocky Mountain Region

Colonel Smith joined the Civil Air Patrol in August 1960, and became a senior member in 1971. Over the years Colonel Smith held positions of Squadron director of Cadets, Administration Personnel, Wing Commander, Wing Administrative Senior Training Chief of Staff, Region Deputy Commander Region Administrator and served as Encampment Commandre in 1971, 1976, 1978 and 1980. In her current post she oversees Civil Air Patrol Corporate activity in five states: Colorado, Idaho, Montana, Utah and Wyoming.

During her more than 30-year tenure with the Civil Air Patrol Virginia Smith has attended numerous region and wing conferences and in 1992 was the project officer for the national board meeting. See fig. 12-3.

Colonel Smith's honors and decorations include the Distinguished Service Award, Exceptional Service Award, Meritorious Service Award, Gill Robb Wilson Award, National Commanders Citation Award, Paul E. Garber Award, and the Grover Loening Award.

Smith, who also has three daughters, is an FCC licensed radio operator, and the president of the American Business Women's Club.

Fig. 12-3 Col. Virginia Smith

Bibliography

Books
- Douglas, Deborah. *United States Women in Aviation 1941-1985* Washington D.C. Smithsonian Studies in Air and Space. 1991.
- Freeman, Lucy. Halbert, Alma. *America's First Woman Warrior: The Courage of Deborah Sampson.* New York, Paragon 1992.
- *Fighter Command,* edited by Jeffery Ethell and Robert T. Sand. 1991 Motorbooks International.
- Griffith, Lynne. McCann, Kelly. *The Book of Women:* 300 Notable Women History Passed By. 1992 Bob Adams, MA.
- Keil, Sally Van Wagenen. *Those Wonderful Women In Their Flying Machines.* New York. Rawson, Wade Publishers, Inc. 1979.
- Lynch, Dr. Barbara, editor. *Proceedings of the Second National Conference on Women in Aviation.* March 21-23 1991.
- Lynch, Dr. Barbara, editor. *Proceedings of the Third National Conference on Women in Aviation.* March 12-14 1992.
- MacDonald, Anne L. *Feminine Ingenuity:* Women and Invention in America. New York. Ballantine Books. 1992.
- Morrissey, Muriel Earhart. Osborne, Carol. *Amelia My Corageous Sister* Santa Clara, Osborne Publisher. 1987.
- Murray, R.C. *Golden Knights - The History of the U.S. Parachute Team.* Canton, Ohio. Daring Books.
- Neprud, Robert E. *Flying Minute Men - The Story of the Civil Air Partol.* New York. Duel Sloan & Pearce 1948.
- Oaks, Claudia. *United States Women in Aviation through World War* I. Smithsonian Studies in Air and Space. Washington, D.C.
- Oaks, Claudia. *United States Women in Aviation 1930-1939.* Smithsonian Studies in Air and Space #6 Washington, D.C.
- Parrish, Berta, Dr. Editor Images of Women in Aviation, *Proceedings from Women in Aviation Conference* March 30, 31 1990.
- Pazmany, Kathleen Brooks *United States Women in Aviation 1919-1929* Smithsonian Studies in Air and Space. Washington, D.C.
- Planck, Charles E. *Women With Wings.* New York, Harper Brothers. 1942.
- Putnam, George Palmer. *Soaring Wings.* New York: Harcourt, 1939.
- Skelton, Betty. *Little Stinker.* 1977. Winter Haven, Cross Press.
- Tate, Grover Ted. *The Lady Who Tamed Pegasus.* Bend,Or. Maverick Publications. 1984.
- Thadden Louise. *High Wide and Frightened.* New York. Stackpole & Sons, 1938.

- Weiser, Marjorie P.K.. Arbeiter, Jean S. *Womanlist.* New York. Atheneum 1981.
- Yeager, Jeana. Rutan, Dick. *Voyager* New York. Alfred Knoff 1987.
- *Past and Promise: Lives of New Jersey Women.* Scarecrow Press 1990. Metchuen, N.J Women's Project of New Jersey.

Magazines
- Ackermann, Joan "A Champion Who's Jumping For A Living." *Sports Illustrated.* December 24, 1991.
- Baker, B. Kimball. "Uncle Sam's Nieces." *Aviation Quarterly* Vol. 7, #3. Second Quarter, 1984.
- Ball, Daisy Elizabeth. "Women's Part in Aviation." *Aeronautical Digest* July 1923.
- Bender, William. "Up and Away" *Hartford Monthly.* March 1989.
- Brown, Margery. "What Men Flyers Think of Women Pilots." *Popular Aviation and Aeronautics.* March 1929.
- Clancy, Frank. *American Health.* "The Flight Stuff." Jan/Feb. 1992.
- Fishman, David J. "Should Women Be Fighter Pilots?" *Science World.* Sept. 20, 1991.
- Floden, Britt. "Just think of the Possibilities." *Soaring* Feb. 1981.
- Garrison, Gene. "1st Lt. Ellen Ausman - Back From the Persian Gulf War." *Carefree Enterprises.* June 1991.
- Goodman, Eric. "Would Their Romance Fly? *TV Guide* 10/22/88.
- Goyer, Norm. "Outside Loop Record" *Sport Pilot* Dec. 1989.
- Hart, Jane "Women in Orbit" *Town and Country* Nov. 1962.
- Jones, Hank "50 Years of the CAP." *Retired Officers Magazine.* January 1992.
- Jones, Terry Gwynn. "For A Brief Moment The World Seemed Wild About Harriet." *Smithsonian Magazine, January, 1984.*
- Klym, Julie Opell. "America's First Flight Academy." *AOPA Pilot.* Oct. 1979.
- Kuhns, Don. "Pancho Barnes: A Legend in Her Own Time." *Virginia Aviation.* Oct.-Dec. 1980.
- Luce, Clare Booth. "Some People Simply Never Get the Message." *Life* Jan. 28, 1963.
- McCullough, Joan. "The Thirteen Who Were Left Behind." *MS* Sept. 1963.
- Moses, Sam. "Sky Princess Passes Her Scepter." *Sports Illustrated* Dec. 18. 1989.
- Ottley, William. "Absolute Accolades." *Parachutist Magazine* December, 1991.

- Orr, Flora. "Southern Personalities." *Holland's the Magazine of the South* September, 1935.
- Pateman, Yvonne C. "WASPs And WAFS in a Fortress." *Aviation Quarterly* Vol. 8, No. 2 Second Quarter 1988.
------- "Rugged and Right - That's Teresa James." *Aviation Quarterly* Vol. 9, No. 1. 1998.
------- "The Story of Ten Grand." *Aviation Quarterly* Vol. 9, No. 1.
--------"Faye Gillis Wells - Aviation Pioneer" *Aviation Quarterly* Vol. 8, No. 1. 1985.
- Phipps, Walter H. The Dangers of the Lifting Tail. *Aircraft* Magazine. August, 1912.
- Quimby, Harriet "How I Made My First Big Flight Abroad." *FLY Magazine.* June, 1912.
- Rickman, Sarah "Cockpit To Classroom - Nagle's in Command." *Women in Aviation The Publication.* August, 1990.
- Vail, Betty. Edwards, Dixon. "Jerrie Mock - Winner Take All." *Flying* July, 1964.
- *Newsweek.* March 30, 1964 "Shades of Amelia."
- *Look Magazine* "The WAFS" February 9, 1943. Winter 1998.
- Now A Civil Air Patrol." *Independent Woman.* January 21, 1942.
- "She Really is Just Plane Crazy." *Women in Aviation* The Publication May/June 1990.
- Albany (Georgia) Journal "Flying High," Betty McNabb Oct. 3, 1969.

Newspapers
- *The Albany (Georgia) Herald.* Albany Woman Pilots T-33 Jet, Describes Experience. June 3, 1954.
Lakeland Ledger 3/21/91 "Aviation, Automotive Records Land Winter Haven Woman in Walk of Fame."
- *Orlando Sentinel* May 15, 1992.
- New York Times Various issues 1919 - present.
- *Staten Islander* Sept. 7, 1911.
- Rasmussen, Cecilia. *Los Angeles Times* "LA Scene Then and Now. April 20, 1992.
- General Aviation News & Flyer Jan. 1992.
- Newspaper clipping Ethel Dare biographical file, Library, National Air & Space Museum.
- Quimby, Harriet. "How I Won My Aviator's License." *Leslie's Weekly* August 24, 1911.
_____ "How A Woman Learns to Fly." *Leslie's Weekly.* May 25, 1911.

_____"An American Girl's Daring Exploits." *Leslie's Weekly.* May 16, 1912.

- *Los Angeles Herald Examiner,* "Dignity And Pride of Skill" Ava Gutierrez Sept 14, 1975.

- *Press-Telegram,* Dianne Smith. "Women and the War." August 11, 1985.

- *The Grunion Gazette,* "Aviation Pacesetter - Meet Barbara London" August 28, 1980.

Other

- Becraft, Carolyn. "Women in the U.S. Armed Services: The War In The Persian Gulf." Women's Research and Education Institute. 1991.

- Pilot's log Jerrie Mock (NASM)

- *Congressional Record* 6/4/87.

- *Congressional Record* July 17, 1962 "Qualifications for Astronauts."

- *World Air Show News.* Jan/Feb. 1989; May/June, 1989; July 1992

- Video taped discussion "Women in Aviation Conference" International Women's Air and Space Museum Jan. 1990.

- Women of Courage (Video) K.M. Productions 1992.

- NASA biographical data sheet Kathryn Thornton.

- Ninety-Nine News Oct 1991.

-.NASA Magazine winter 1991.

- *Airman.* Dec 1991.

Photo Credits:

Chapter One - Fig. 1-1, 1-2 Smithsonian Institution, Fig. 1-3 author's collection, Fig. 1-4, 5, 6, 7 Smithsonian Institution.

Chapter Two - Fig. 2-1 Smithsonian Institution, Fig. 2-2 Carol Osborne, Fig. 2-3 author's collection, Fig. 2-4 H.V. Pat Reilly, Fig. 2-5 Fay Gillis Wells, Fig. 2-6 Carol Osborne, Fig. 2-7 Smithsonian Institution, Fig. 2-8 International Women's Air and Space Museum.

Chapter Three - Fig. 3-1 Smithsonian Institution, Fig. 3-2 Texas Women's University, Fig. 3-3 H. Glenn Buffington, Fig. 3-4 Texas Women's University, Fig. 3-5 Kay Brick, Fig. 3-6 H.V. Pat Reilly, Fig. 3-7 Barbara Erickson London, Fig. 3-8 Texas Women's University.

Chapter Four - Fig. 4-1 McDonnell Douglas photo by Don Dinkelkamp, via Trish Beckman, Fig. 4-2 Ann Patrie, Fig. 4-3 Lori Cone, Fig. 4-4 Kelly Franke, Fig. 4-5 Ellen Ausman, Fig. 4-6 Sue Patelmo, Fig. 4-7 Debbie Myers, Fig. 4-8 Lucy Young, Fig. 4-9 Lisa Williams, Fig. 4-10 Carole Danis Litten.

Chapter Five - Fig. 5-1, 2 Betty Skelton, Fig. 5-3, 4 Patty Wagstaff, Fig. 5-5 Joann Osterud, Fig. 5-6 Julie Clark, Fig. 5-7, 8 Marilyn Fitzgerald, Fig. 5-9, 10 Jim Mumaw, Fig. 5-11 Cheryl Stearns, Fig. 5-12, 13 Misty Blues, Fig. 5-14, 15 Jessie Woods, Fig. 5-16 Gene Littlefield Shows, Fig. 5-17, 18 Jim Larson photo via Lori Lynn Ross, Fig. 5-19 Pat Wagner, Fig. 20 Beth Sullivan, Fig. 5-21 Kathy Wadsworth, Fig. 5-22, 23 Britt Floden, Fig. 5-24 Virginia Schweizer.

Chapter Six - Fig. 6-1 Alinda Wikert, Fig. 6-2 Wanda Whitsitt, Fig. 6-3 Ann Gray, Fig. 6-4 Cicele Hatfield, Fig. 6-5 Betty McNabb, Fig. 6-6 Amy Carmien, Fig. 6-7 Racquel McNeil, Fig. 6-8 Moya Lear, Fog. 6-9 Fran Grant.

Chapter Seven - Fig. 7-1 Smithsonian Institution, Fig. 7-2, 3, 4, 5, 6, NASA, Fig. 7-7 Rose Loper, Fig. 7-8 Susan Darcy, Fig. 7-9 Jeana Yeager via Bill Williams.

Chapter Eight - Fig. 8-1 Arlene Feldman, Fig. 8-2 Dayle Buschkotter, Fig. 8-3, 4 Wally Funk, Fig. 8-5 Catherine Roetzler.

Chapter Nine - Fig. 9-1 Peggy Baty, Fig. 9-2 Joan Mace, Fig. 9-3 Jody Pfeifer photo, Fig. 9-4 Carol Osborne, Fig. 9-5 Yvonne Pateman, Fig. 9-6 Doris Lockness, Fig. 9-7 Nancy Hopkins Tier, Fig. 9-8 via Joan Hrubec.

Chapter Ten - Fig. 10-1 H. Glenn Buffington, Fig. 10-2 Susan Maule, Fig. 10-3 Karen kahn, Fig. 10-4, UPS photo via karen Davies Lee, Fig. 10-5 Connie Tobias, Fig. 10-6 Terry London Rinehart, Fig. 10-7 Cyd Laurie Fougner, Fig. 10-8 "Universitas" The Magazine of St. Louis University, via Betsy Carroll Smith.

Chapter Eleven - Fig. 11-1, 2 Mardo Crane, Fig. 11-3, 4, 5, 6, 7 via Kay Brick, Fig. 11-8 via Ann Cooper, Fig. 11-9, 10 Fran Bera, Fig. 11-11 Susan Dusenbury, Fig. 11-12 via Marion Jayne Palazola.

Chapter Twelve - Fig. 12-1 H.V. Pat reilly, Fig. 12-2 Chicago Tribune, Fig. 12-3 Virginia Smith.

Contributors
(Alpha)

Ellen Ausman
Peggy Baty
Trish Beckman
Fran Bera
Janet Harmon Waterford Bragg
Kay Brick
Janis Blackburn
H. Glenn Buffington
Dayle Buschkotter
Amy Carmien
Julie Clark
Eileen Collins
Tucker Comstock
Laurie Cone
Ann Cooper
Mardo Crane
Susan Darcy
Susan Dusenbury
Andy Edmondson
Arlene Feldman
Sharon Fitzgerald
Britt Floden
Cyd Fougner
Kelly Franke
Wally Funk
Anne Gray
Marjorie Gray
Laura Goldsberry
Kelly Hamilton
Cecile Hatfield
Karen Kahn
Moya Lear
Karen Lee
Cheryl Rae Littlefield
Carole Danis Litten
Doris Lockness
Barbara Erickson London
Rose Loper
Joan Mace

Susan Maule
Betty McNabb
Racquel McNeil
Marta Bohm-Meyer
Joan Morris
Katrina Mumaw
Jim Mumaw
Deborah Myers
Carol Osborne
Catherine Osman
Joann Ostrud
Marion Jayne Palozola
Ann Patrie
Sue Patelmo
Yvonne "Pat" Pateman
H.V ."Pat" Reilly
Terry London Rinehart
Catherine Roetzler
Lori Lynn Ross
Virginia Schweizer
Betty Skelton
Sharon Smith
Betsy Carroll Smith
Cheryl Stearns
Beth Sullivan
Pam Kleckner Sullivan
Michelle Tallon
Nancy Hopkins Tier
Connie Tobias
Bobbi Trout
Terry VandenDolder
Kathy Wadsworth
Pat Wagner
Patty Wagstaff
Fay Gillis Wells
Wanda Whitsitt
Alinda Hill Wikert
Lisa Williams
Sandra Williams
Jessie Woods
Jeana Yeager
Lucy Young

RESOURCES:
For further information on aviation careers contact the following
organizations:

Ninety-Nines
Will Rogers Airport
PO Box 59965
Oklahoma City, OK 73159
Loretta Gragg, Executive Director

International Society of Women Airline Pilots
ISA+21
PO Box 66268
Chicago, IL 60666-0268

Whirly-Girls
International Women Helicopter Pilots
PO Box 58484
Houston, TX. 77058
Colleen Nevius, Executive Director

Women Military Aviators
PO Box 396
Randolph Air Force base, TX. 78148
Co. Kelly Hamilton, Pres.

Women in Aerospace
6212-B Old Keene Mill Ct.
Springfield, VA. 22152

Professional Women Controllers, Inc.
PO Box 44085
Oklahoma City, OK. 73144

Women Soaring Pilots Association
c/o Sharon Smith
801 Elsbeth
Dallas, TX. 75208

International Women's Air and Space Museum
26 North Main St.
PO Box 465
Centerville, OH. 45459

Future Aviation Professionals
4959 Massachusetts Ave.
Atlanta, GA. 30337

Women in Aviation - The Publication
PO Box 4323
Traverse City, MI 49685

AirLifeLine
1716 X Street
Sacramento, CA 95818

FAA Headquarters
Education Officer
800 Independence Ave SW
Washington, DC 20591

Aviation Colleges and Universities

Embry-Riddle Aeronautical University
600 S. Clyde Morris Blvd.
Daytona Beach, FL. 32144

Embry-Riddle Aeronautical University
3200 N. Willow Creek Rd.
Prescott, AZ 86301

Parks College of Saint Louis University
Cahokia, IL. 62206

Florida Institute of Technology
150 West University Blvd.
Melbourne, FL 32901

Indiana State University
West Lafayette, IN 47907

Middle Tennessee State Univ
PO Box 67 MTSU
Murfreesboro TN 37132

Reference Notes

Chapter One
1- Marjorie Weiser. Jean Arbeiter. *Womanlist*. Atheneum New York 1981 p.163
2- Anne L. MacDonald. *Feminine Ingenuity: Women and Invention in America*. Ballantine Books. 1992
3- Ibid
4- Charles E. Planck. *Women With Wings*. New York, Harper Brothers. 1942 p.14
5- Congressional Record June 4, 1987
6- Julie Opell Klym. *AOPA Pilot*. Oct. 1979. "America's First Flight Academy".
7- Harriet Quimby. "How I Won My Aviator's License." *Leslie's Weekly* August 24,1911.
8- Harriet Quimby. "How A Woman Learns to Fly." *Leslie's Weekly*. May 25, 1911.
9- New York *Times* August 2, 1912.
10- Matilde Moisant's Early Bird application.
11- Told to Henry Holden by Bobbi Trout July 1992.
12- "Harriet Quimby. How A Woman Learns To Fly." *Leslie's Weekly*, May 25, 1911.
13- *Staten Islander* Sept. 7, 1911.
14- Harriet Quimby. "An American Girl's Daring Exploits." *Leslie's Weekly*. May 16, 1912.
15- Ibid.
16- Op. cit.
17- Op. cit.
18- *FLY* Magazine, June 1912.
19- Harriet Quimby. "An American Girl's Daring Exploits." *Leslie's Weekly*. May 16, 1912.
20- Ibid.
21- Op. cit.
22- Op. cit.
23- Photo evidence NASM, Smithsonian Institution.

24- Walter H. Phipps. The Dangers of the Lifting Tail. *Aircraft Magazine*. August 8, 1912.

25- Claudia Oakes. *United States Women in Aviation through World War I* Smithsonian Studies in Air and Space. p.22.

26- Ibid.

27- Charles E. Planck. *Women With Wings*. New York, Harper Brothers. 1942 p.37.

28- New York *Times* October 22, 1919 p.22.

29- Kathleen Brooks Pazmany. *United States Women in Aviation 1919-1929*. Smithsonian Studies in Air and Space p.4.

30- Ibid p.6.

Chapter Two

1- Flora Orr. "Southern Personalities." *Holland's the Magazine of the South* September, 1935.

2- Kathleen Brooks-Pazmany *United States Women in Aviation 1919-1929* Smithsonian Studies in Air and Space p.2.

3- H.V. Pat Reilly. *From the Balloon to the Moon*. New Jersey. H.V. Publishers. p.125

4- 1970 International Convention. Heritage Night Dinner Program.

5- George Palmer Putnam. *Soaring Wings*. New York: Harcourt, 1939 p.79.

6- Muriel Earhart Morrissey. Carol Osborne. *Amelia My Courageous Sister*. Santa Clara, Osborne Publisher 1987. p.119.

7- Dr. Barbara Lynch, editor. *Proceedings of the Second National Conference on Women in Aviation*. March 21-23 1991. p61.

8- Muriel Earhart Morrissey, Carol Osborne. *Amelia My Courageous. Sister* Santa Clara, Osborne Publisher 1987 p.102.

9- Dr. Barbara Lynch, editor. *Proceedings of the Second National Conference on Women in Aviation*. March 21-23 1991 p.62.

10- Aviation Quarterly Vol. 8, No. 1 1985.

11- Don Kuhns. *Virginia Aviation* "Pancho Barnes: A Legend in Her Own Time." Oct.-Dec. 1980 p.6.

12- Cecilia Rasmussen. Los Angeles *Times* "LA Scene Then and Now." April 20, 1992.

13- Don Kuhns. *Virginia Aviation* "Pancho Barnes: A Legend in Her Own Time." Oct.-Dec. 1980 p.6.

14- Cecilia Rasmussen. Los Angeles *Times* "LA Scene Then and Now." April 20, 1992.

15- Eric Goodman. "Would Their Romance Fly?" *TV Guide* October 22, 1988.

16- Ibid

17- Cecilia Rasmussen. Los Angeles *Times* "LA Scene Then and Now. "April 20, 1992.

18- Eric Goodman. "Would Their Romance Fly?" *TV Guide* October 22, 1988.

Chapter Three

1- The *Proceedings From Images of Women In Aviation*: *Fact vs. Fiction Conference:* p.21.

2- Deborah G. Douglas. *United States Women in Aviation 1940-1985*. Smithsonian Institution Press p.vi.

3- Ibid p.17.

4- Op. cit.

5- Op. cit.

6- The *Proceedings From Images of Women In Aviation*: *Fact vs. Fiction Conference*: p.1.

7- Lynne Griffith, Kelly McCann. *The Book of Women: 300 Notable Women History Passed*. By. Bob Adams, MA. 1992. p.32.

8- *The Proceedings From Images of Women In Aviation*: *Fact vs. Fiction Conference:* p.1.

9- "Now A Civil Air Patrol." *Independent Woman*. January 21, 1942 p.44.

10- *Past and Promise: Lives of New Jersey Women*. Scarecrow Press 1990. p.400.

11- See *Ladybirds - The Untold Story of Women Pilots in America*.

12- *United States Women in Aviation 1940-1985*, Smithsonian Institution Press 1985 p.44 .

13- Letter from Nancy Love to J.C. Bradford Beech Aircraft Corp. 5/24/55 NASM biographical files of Nancy Love.

14- Deborah Douglas. *United States Women in Aviation 1941-1985* Washington D.C. Smithsonian Studies in Air and Space. 1991. p.29.

15- Ibid p.30.

16- Ibid. p.42.

17-Ibid. p.46.

18- See *Ladybirds - The Untold Story of Women Pilots in America*.

19- Letter from Nancy Love to J.C. Bradford Beech Aircraft Corp. 5/24/55 NASM biographical files of Nancy Love.

20- Ibid.

21- B. Kimball Baker, "Uncle Sam's Nieces." *Aviation Quarterly* Vol. 7, #3. p.250.

22- *Fighter Command*, edited by Jeffery Ethell and Robert T. Sand. Motorbooks International. 1991.

23- *Women of Courage* video. KM Productions. Lakewood, CO

24- note to author.

25- Dianne Smith "Women and the War." *Press-Telegram* Sunday August 11, 1985.

26- *Airman* Magazine of America's Air Force Dec. 1991 pg 33.

27- "Dignity And Pride of Skill" Ava Gutierrez. *Los Angeles Herald Examiner* Sept 14, 1975.

28- Marianne Verges. *On Silver Wings* Balantine Books. 1991. pg.137.

29- *Women of Courage* video. KM Productions. Lakewood, CO.

30- "Aviation Pacesetter - Meet Barbara London" *The Grunion Gazette.* August 28, 1980

31- "The WASPS In Relation To The Army Air Forces, Prepared for: Director of Operations and Training by The Historical Officer" December 7, 1944.

32- Women in Aviation Conference IWASM 1990.

33- Ibid.

34- *Women of Courage* video. KM Productions. Lakewood, Co.

35- Ibid.

36- Op.cit.

37- *The Proceedings From Images of Women In Aviation: Fact vs. Fiction Conference:* p.5.

38- Byrd Howell Granger. *On Final Approach.* Scottsdale, AZ. Falconer Publishing. 1991. p.250.

39- General Arnold's closing remarks WASP graduation December 7, 1944.

40- Lynne Griffith, Kelly McCann. *The Book of Women 300 Notable Women History Passed By.* Bob Adams, MA. 1992. p.106.

41- *Woman of Courage* video. KM Productions. Lakewood, CO.

42- Deborah Douglas. *United States Women in Aviation 1941-1985.* Washington D.C. Smithsonian Studies in Air and Space. 1991. p.55.

43- Sally Van Wagenen Keil. *Those Wonderful Women In Their Flying Machines.* New York Rawson, Wade Publishers, Inc. 1979. p.306.

44- *Flying,* Dec. 1944.

45- *Women of Courage* video. KM Productions. Lakewood, CO.

46- Ibid.

Chapter Four

1- Lynne Griffith, Kelly McCann. *The Book of Women: 300 Notable Women History Passed* By. Bob Adams, MA. 1992. p.58.

2- ibid p.51.

3- ibid p.45.

4- Louise Thadden. *High Wide and Frightened.* Stackpole & Sons, 1938 p.239.

5- Deborah Douglas. *United States Women in Aviation 1941-1985.* Washington D.C. Smithsonian Studies in Air and Space. 1991 p.27.

6- ibid p. 91.

7- Lynne Griffith, Kelly McCann. *The Book of Women: 300 Notable Women History Passed By.* Bob Adams, MA. 1992. p. 89.

8- Helen Collins. *Naval Aviation News.* "From Plane Captains to Pilots." July 1977.

9- Deborah Douglas. *United States Women in Aviation 1941-1985.* Washington D.C. Smithsonian Studies in Air and Space. 1991 p.102.

10- Frank Clancy, "Breaking the Sexbarrier." *American Health.* January/February 1992.

11- Ibid.

12- See *Ladybirds - The Untold Story of Women Pilots in America.* Black Hawk Publishing Co.

13- David J. Fishman. *Science World* "Should Women Be Fighter Pilots? Sept. 20, 1991.

14- *USA Today* 5/6/92.

15- Carolyn Becraft. "Women in the U.S. Armed Services: The War In The Persian Gulf." Women's Research and Education Institute. 1991.

16- Ibid.

17- Ibid.

Chapter Five

1- International Council of Air Shows.

2- Betty Skelton. *Little Stinker.* Winter Haven, Cross Press, 1977. p.23.

3- ibid p.19.

4- International Aerobatic Club.

5- *Sports Illustrated* December 18, 1989.

6- Ibid.

7- *Ninety-Nine News.* Oct. 1991.

8- *World Air Show News* J-F 1989.

9- *Sports Illustrated.* December 24, 1990 P.16.

10- Ibid.

11- Ibid. P.17.

12- *Golden Knights.* History of the U.S. Army Parachute Team. p.12.

13- *Sports Illustrated.* Dec. 24, 1990. p.17

14- *Women in Aviation* May-June 1989 Vol 1, Issue 2 pg. 3.

15- *World Airshow News* May/June 1989.

16- Paul Maravelas. *Ballooning.* Interview with Jeannette Piccard. April 17, 1980.

17- Ibid.

18- *Past and Promise: Lives of New Jersey Women.* Scarecrow Press Metchuen, N.J 1990. Women's Project of New Jersey.

19- Deborah Douglas. *United States Women in Aviation 1941-1985.* Washington D.C. Smithsonian Studies in Air and Space. 1991 p.55.

20- *Past and Promise: Lives of New Jersey Women.* Scarecrow Press Metchuen, N.J 1990. Women's Project of New Jersey.

21- Lynch, Dr. Barbara, editor. *Proceedings of the Third National Conference on Women in Aviation.* March 12-14 1992. p.70

Chapter Seven

1- Joan Merriam Smith. "I flew Around the World Alone." *Saturday Evening Post.* July 25 - August 1, 1964. p. 77-83.

2- *Flying* July 1964 p.66.

3- Ibid p.60.

4- Ibid p.56.

5- James Fulton, *Congressional Record* July 18, 1962. p.57.

6- Deborah Douglas *United States Women in Aviation 1941-1985.* Washington D.C. Smithsonian Studies in Air and Space. 1991.

7- *Congressional Record* July 18, 1962.

8- Ibid.

9- Ibid p.48.

10- Ibid p.13.

11- Ibid p.41.

12- Ibid p.56.

13- Ibid p.17.

14-Ibid p.22.

15- Frank Robinson, "Conversation With Robert Heinlein."

16- *Orlando Sentinel* May 15, 1992.

17- Ibid.

18- *NASA Magazine* winter 1991.

19- Ibid.

20- *Plane & Pilot* "An Officer and a Gentlewoman" July, 1992. p.31.

21- Ibid.

22- *Voyager* Jeana Yeager and Dick Rutan. Knoff 1987.

Chapter Eight

1- Deborah Douglas US *Women in Aviation 1940-1985.* Smithsonian Air and Space Studies.

INDEX